ENGINEERING TOOLS, TECHNIQUES AND TABLES

ULTRASONICS

THEORY, TECHNIQUES AND PRACTICAL APPLICATIONS

ENGINEERING TOOLS, TECHNIQUES AND TABLES

Additional books in this series can be found on Nova's website
under the Series tab.

Additional e-books in this series can be found on Nova's website
under the e-book tab.

MECHANICAL ENGINEERING THEORY AND APPLICATIONS

Additional books in this series can be found on Nova's website
under the Series tab.

Additional e-books in this series can be found on Nova's website
under the e-book tab.

ENGINEERING TOOLS, TECHNIQUES AND TABLES

ULTRASONICS

THEORY, TECHNIQUES AND PRACTICAL APPLICATIONS

HANAKO AYABITO
AND
MITSUKO KATSUKAWA
EDITORS

nova publishers
New York

For permission to use material from this book please contact us:
Telephone 631-231-7269; Fax 631-231-8175
Web Site: http://www.novapublishers.com

NOTICE TO THE READER

Additional color graphics may be available in the e-book version of this book.

LIBRARY OF CONGRESS CATALOGING-IN-PUBLICATION DATA

Ultrasonics : theory, techniques and practical applications / editors, Hanako Ayabito and Mitsuko Katsukawa.
p. cm.
Includes index.
ISBN: 978-1-62257-685-2 (hardcover)
1. Ultrasonics--Research. I. Ayabito, Hanako. II. Katsukawa, Mitsuko.
TA367.U48 2011
320.2'8--dc23 2012030289

Published by Nova Science Publishers, Inc. † New York

CONTENTS

PREFACE

Ultrasonics has been recognized not only as a viable and dependable tool to prognosticate diseases in medical practices but also a prevailing technique in non-destructive detection and manufacturing. In this book, the authors study the theory, techniques and practical applications of ultrasonics. Topics include mechanochemical cleavage of DNA by ultrasound; using ultrasonic wave transmission and wave reflection methods to monitor the setting and hardening process of cement based materials; ultrasonic spray-pyrolysis process for materials processing; ultrasonic assisted fixed abrasive machining of hard-brittle materials; dynamics and kinetics of supercooled liquids under ultrasonic field; mathematical model of a traveling wave ultrasonic motor; ultrasonic array image in anisotropic weld structures; and ultrasonics in machining operations.

Chapter 1 - Ultrasonic cleavage of polymers has been a subject of intensive studies during the past decades. Currently used experimental protocols and techniques of nucleic acids manipulation allow us to conduct ultrasonic cleavage experiments with DNA fragments of known length and base pair sequence. Further analysis of ultrasonic cleavage products with polyacrylamide gel electrophoresis makes it possible to determine the exact positions of DNA cleavage sites and corresponding cleavage efficiencies. In the authors' s study they analyze the phenomenon of ultrasonic cleavage of DNA in details. Two main effects are observed in DNA cleavage patterns: attenuation of ultrasonic cleavage at sites close to the ends of the DNA fragments and sequence-specificity of cleavage. Here the authors propose the model which accounts for these two effects. The authors' model describes the extension of the DNA molecules in cavitational flow that leads to the mechanochemical cleavage of DNA. The sequence specificity of ultrasonic cleavage is shown to be likely the result of sequence-dependent conformational dynamics. Hence, the relative efficiency of ultrasonic cleavage may serve as indicator of sequence specific flexibility in both strands of DNA. Such numerical evaluation may be useful for identifying promoter regions in genomes and for assessing the preference for nucleosome positioning.

Chapter 2 - During the recent years comprehensive studies have been performed focusing on the possibility of using different ultrasonic techniques to monitor a hydration process and formation of structure of cement based materials at early ages. As a result, researchers all over the world have published many scientific articles dealing with this topic.

Among these, ultrasonic wave transmission and ultrasonic wave reflection methods are generally used. Using these techniques, the hydration process of various cementitious materials with different compositions can be measured continuously and non-destructively.

Moreover, the methods have been proven to be very sensitive to the presence of different chemical and mineral admixtures in the material composition. On the basis of different ultrasonic parameters, procedures have been proposed to identify important milestones that appear during the formation of structure of the material, including initial and final setting time. Ultrasonic parameters have been successfully correlated to various material properties, namely the percolation of solid phases, compressive strength, elastic and shear modulus, degree of cement hydration, degree of capillary porosity, viscosity etc. Undoubtedly, these results improve practical application of the ultrasonic methods to a high degree.

In this chapter, an overview of some most important results dealing with the possibility of using ultrasonic wave transmission and ultrasonic wave reflection methods to monitor the setting and hardening process of cement based materials at early ages is presented.

Chapter 3 - Ultrasonic spray-pyrolysis process has been utilized for the past forty years with great convenience for materials processing, especially for the fabrication of metal oxide thin films of semiconducting nature. The process renders several advantages associated to the adhesion, control of morphology, particle size, and homogeneity of thin films, apart from its low cost.

In this brief review, the authors present a short description of ultrasonic nebulization process of liquid through ultrasonic wave induced cavitation. Effects of different parameters like the ultrasound frequency, ultrasonic power, and fluid level in the atomization chamber on the nebulization rate of a liquid solvent of precursor have been discussed.

Chapter 4 - Ultrasonics has been identified as a practical and vital technique in manufacturing, especially in the high efficiency and precision machining of difficult-to-machine materials, which, in some cases, is otherwise considerably difficult to be accomplished without ultrasonics. Traditional machining in the presence of ultrasonics can be categorized into ultrasonic machining (USM) and ultrasonic assisted machining (UAM). The USM involves the use of ultrasonics to energize loose abrasives and the energized abrasives impinge upon surfaces of workpieces to remove material in the form of mechanical actions such as hammering, rolling and sliding of abrasives and occasionally chemical actions-ultrasonic cavitation erosion. In contrast with USM, UAM refers to machining that ultrasonics is incorporated into widespread machining techniques such as cutting and grinding. The UAM has proven to enhance the machinability of difficult-to-machine materials and/or improve the quality of workpiece surfaces, which is due greatly to exceptional ability to dispel machining debris and/or to reduce machining forces and resultant machining heating.

This chapter concentrates on the introduction of the UAM technique where ultrasonics is applied to fixed abrasive machining as there is an increasing tendency to supersede loose abrasives owing mainly to low-efficiency and cost-ineffectiveness of loose abrasive machining. The chapter commences with a succinct introduction of the principles of the USM and the UAM and then proceeds to internal grinding with the aid of ultrasonics, which is followed by ultrasonic assisted grinding of mono-crystalline silicon and sapphire, and ends with a recently proposed technique of ultrasonic assisted fixed abrasive polishing of optical glass. In each part, processing principle and apparatus will be described concisely and the machining methods are ultimately evaluated in terms of surface roughness and/or material removal rate, etc.

Chapter 5 - In simple system (metallic sodium, benzene and naphthalene), supercooled states were investigated by ultrasonic measurements. Frequency dependent shear loss modulus can detect relaxation process sensitively in the supercooled liquid of metallic sodium. The dynamic heterogeneity might correspond to a dynamic cluster at the supercooled state. In contrast to the monatomic system, benzene and naphthalene possess anisotropic interactions intrinsically due to anisotropic molecular shapes.

The peculiar feature is characterized by a competition between π-π repulsive interaction and molecular packing efficiency. On balancing in the local circumstance, supercooled benzene is coupled with the ultrasonic fields. Furthermore, the ultrasonic irradiation effect is enhanced by thermal treatments such as quenching. Kinetics during a non-equilibrium state can extract hidden information about the nucleation process, where dynamic heterogeneous fluctuations in the supercooled liquids are dominant in dissipative system.

Chapter 6 - The traveling wave ultrasonic motor has excellent characteristics. It has advanced features such as high-holding torque without supply high torque at low speed, absence of parasitic magnetic fields, continuous contact, high response, quiet operation and small size. All these properties make of this motor an ideal actuator, comparatively with the electromagnetic one. During the last decade, many prototypes of ultrasonic motor have been designed as a function needs. The advanced features are due to the working principle that is based on high-frequency vibrations and due to the tribological phenomenon acting in the zone of contact between stator and rotor. There is not a perfect contact model that can control the contact dynamic between the rotor and the stator because of it has a nonlinear behavior. For this reason, almost all research carried out is based on a requirement that leads to some errors between the experimental and theoretical results. Modeling of traveling wave ultrasonic motors (TWUM), has been, so, subject much research.

In this chapter, optimal deformation angle and resonance frequency of frictional material are investigated. Force expressions in the stator/rotor interface are developed. Mechanical characteristics taking into account the effect of shearing deformation are deduced and presented. Results of the simulation are validated using the measurements made by other researchers on traveling wave piezoelectric motor AWM90-X, Daimler-Benz type, and a prototype USM (AWM-90).

Chapter 7 - The quality of an ultrasonic array image depends on accurate information about its acoustic properties. Inaccuracy of acoustic properties can cause image degradation such as blurring, mislocation of reflectors and the introduction of artefacts. In this chapter, for the specific case of an inhomogeneous and anisotropic austenitic steel weld, Monte Carlo Markov Chain (MCMC) inversion is used to estimate unknown acoustic properties from array data. The approach uses active beacons that transmit ultrasound through the anisotropic weld which is captured by a receiving array. A forward model of the ultrasonic array data is then optimised with respect to the experimental data using a MCMC inversion. The result of this process is the extraction of a material property map that describes the anisotropy distribution within the weld region. These extracted material properties are then used within an imaging algorithm, the Total Focusing Method, to produce autofocused images. This MCMC inversion approach is first applied to simulated data which is used to test the convergence, robustness and accuracy of the method and its implementation. In this case the extracted weld map is used to show improved imaging of defects within the weld as compared to an image formed assuming a constant velocity.

Chapter 8 - The objective of the vibration assisted machining techniques is to improve the traditional cutting processes by the application of an external vibration to the cutting edge to achieve improvement in the chip removal mechanism and in the manufactured part quality.

Currently, two techniques are most common for vibration-assisted machining due to their feasibility and industrial expectations: the Modulation-Assisted Machining and the Ultrasonic Assisted Machining.

Modulation-Assisted Machining superimposes a controlled low-frequency oscillation to the cutting process, typically up to 1000Hz with amplitudes up to 500 micrometers. The mechanics of the chip formation is changed in that way, and cutting is divided into discrete events by means of control of the modulation parameters. Thus, it improves efficiency and process capability in precision machining processes.

On the other hand, Ultrasonic Assisted Machining consists in adding high frequency excitation, which can reach 40 KHz, and low vibration amplitude, ranging between 3 and 30 micrometers, to the tool or workpiece to improve the cutting process. Ultrasonic assisted machining technology demonstrates a range of benefits in both i) high precision machining applications and ii) machining of difficult-to-cut alloys and brittle materials. Ultrasonic assistance is applied to several cutting processes, but most of the research on this method is focused on turning and drilling processes.

In this chapter, the authors report the basis of these techniques and the advantages and drawbacks of ultrasonic assisted turning. Thus, the effect of tool vibration on cutting process and workpiece surface quality was studied in turning of mild steel. Several parameters of ultrasonic assisted turning (UAT) were monitored, including surface roughness, tool wear, chip formation and tool temperature. The results obtained show that ultrasonic vibration can improve the surface quality decreasing surface roughness and cutting tool temperature and also a tool life increase is achieved compared to conventional turning (CT).

In: Ultrasonics: Theory, Techniques and Practical Applications ISBN: 978-1-62257-685-2
Editors: Hanako Ayabito and Mitsuko Katsukawa

Chapter 1

MECHANOCHEMICAL CLEAVAGE OF DNA BY ULTRASOUND

***Sergei L. Grokhovsky*[1], *Irina A. Il'icheva*[1], *Dmitry Yu. Nechipurenko*[2], *Mikhail V. Golovkin*[1] *Larisa A. Panchenko*[3], *Robert V. Polozov*[4] and *Yury D. Nechipurenko*[1]**

[1]Engelhardt Institute of Molecular Biology,
Russian Academy of Sciences, Moscow, Russia
[2]Department of Physics, Moscow State University, Moscow, Russia
[3]Department of Biology, Moscow State University, Moscow, Russia
[4]Institute of Theoretical and Experimental Biophysics, Russian Academy of Sciences, Puschino, Moscow Region, Russia

ABSTRACT

Ultrasonic cleavage of polymers has been a subject of intensive studies during the past decades. Currently used experimental protocols and techniques of nucleic acids manipulation allow us to conduct ultrasonic cleavage experiments with DNA fragments of known length and base pair sequence. Further analysis of ultrasonic cleavage products with polyacrylamide gel electrophoresis makes it possible to determine the exact positions of DNA cleavage sites and corresponding cleavage efficiencies. In our study we analyze the phenomenon of ultrasonic cleavage of DNA in details. Two main effects are observed in DNA cleavage patterns: attenuation of ultrasonic cleavage at sites close to the ends of the DNA fragments and sequence-specificity of cleavage. Here we propose the model which accounts for these two effects. Our model describes the extension of the DNA molecules in cavitational flow that leads to the mechanochemical cleavage of DNA. The sequence specificity of ultrasonic cleavage is shown to be likely the result of sequence-dependent conformational dynamics. Hence, the relative efficiency of ultrasonic cleavage may serve as indicator of sequence specific flexibility in both strands of DNA. Such numerical evaluation may be useful for identifying promoter regions in genomes and for assessing the preference for nucleosome positioning.

1. Introduction

The earliest observations of DNA ultrasonic fragmentation in aqueous solution were reported in 1952 (Laland et al., 1952). It was found that the viscosity of DNA solutions decreased with exposure to sonic waves, whereas the optical density at 260 nm remained unchanged. Subsequent analyses have shown that ultrasound irradiation of DNA solution leads to the cleavage and formation of double-stranded oligonucleotides of various lengths with a phosphorylated 5'-end and a hydroxyl group at the 3'-end (Doty et al., 1958; Davison et al., 1962). Polymeric chains break mostly under the action of extremely large forces arising in rapid flows near collapsing cavitation bubbles and shock waves arising as the result of violent bubbles collapse. Sonication of double-stranded DNA in an aqueous solution causes mostly double-strand breaks and yields fragments of several hundreds of base pairs (Pritchard et al., 1965; Elsner et al., 1989). Recently we showed that DNA ultrasonic cleavage may be characterized by two major effects: the positional effect (i.e. the dependence of phosphodiester bond cleavage efficiency on the position of the bond along DNA fragment) and sequence-specificity of cleavage (the dependence of cleavage pattern on DNA base pair sequence).

Experiments have shown that ultrasonic cleavage of DNA represents a mechanochemical reaction induced by cavitational processes in irradiated solution (Basedow et al., 1977). Recently using polyacrylamide gel electrophoresis we studied solutions containing 3'-end-labeled double-stranded DNA fragments treated with 22 kHz or 44 kHz ultrasound and found that the intensity of cleavage at a given position depends on the nucleotide sequence and occurs predominantly at the 5'-CpG-3' dinucleotide (Grokhovsky, 2006; Grokhovsky et al., 2008).

2. Methods and Experimental Protocols

Restriction fragments of DNA were generated by digestion of plasmid DNA by the restriction endonucleases. The fragments were 3'-end-labeled with [α-^{32}P]dATP in the presence of the unlabeled other dNTP and the Klenow fragment of *E. coli* DNA-polymerase I. The DNA fragments were isolated by nondenaturing polyacrylamide gel in a 1-mm thick 5% gel with subsequent elution and precipitation (Maniatis et al., 1982).

For sonication, 10 mkL of DNA fragments (~10^4 Bq) in water were mixed with 10 mkL of 0.2M NaOAc in the bottom of a thin-walled polypropylene microcentrifuge tube of 0.2 mL capacity. The final concentration of the fragments was 5–10 mkg/mL (~10 mkM base pair). The test tubes were placed in a Teflon ring that contained a central slot (15 mm in diameter) and radial slots for the test tubes. The test-tube ends were located ~0.5 cm below the horn sonicator edge, which had a diameter of 12 mm. The ring and horn sonicator were placed in a vessel with water and crushed ice. Ultrasound was generated by a 300 W generator with a frequency of 44 or 22 kHz using the maximum power output. The ultrasonic power exerted on the system was determined calorimetrically and exceeded 60 W. To obtain rough measurements of the chemical effects induced by cavitation inside the tubes, we used a test tube containing 0.05 M KI in 0.025% starch solution. After 8 min of irradiation, the coloration extent was equal to that obtained by adding ~0.1 mM of hydrogen peroxide. This yield is comparable to reported results obtained under normal temperature conditions and with ultrasound intensity exceeding 2 W/cm2 (Margulis, 1984). It is worth noting that both low ultrasound frequency and low temperature conditions are known to increase the power of cavitational effects (Basedow & Ebert, 1977).

Figure 1 demonstrates cleavage profiles of the 470-bp DNA fragment in nondenaturing gel. After sonication the samples were combined with an equal volume of 50% glycerol with 0.02% bromphenol blue. Aliquots (0.5 mkl) were applied on polyacrylamide gel of 40 cm in length and 0.15 mm in thickness. Electrophoresis was carried out in TBE at 1.3 kV (~30W) for 3 h. Gels were dried on a glass which was pretreated with γ-methacrylpropyloxysilane and consequently exposed with a luminescent screen. The fragment contained AT clusters alternating with GC clusters. Analysis of the nucleotide sequence of these and other fragments demonstrated that DNA strands break more readily between cytosine and guanine in the 5'-CpG-3' sequence (Grokhovsky, 2006). This fragment was cleaved preferentially at several sites, which corresponded to alternating GC pairs.

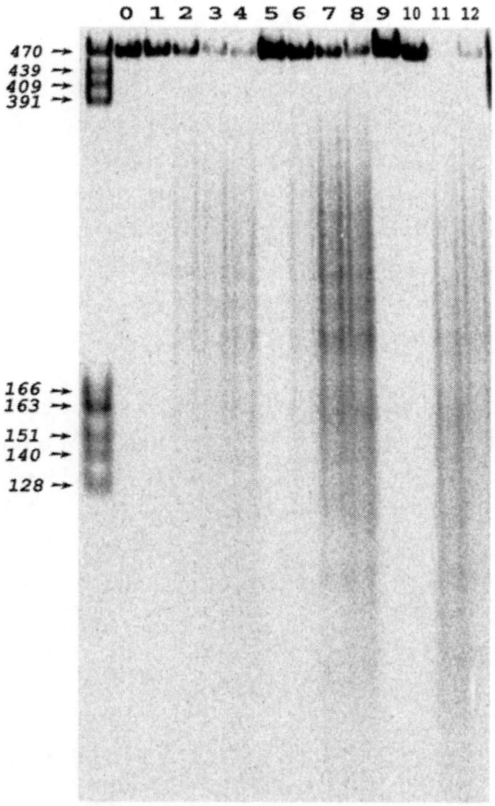

Figure 1. Cleavage profiles of the 470-bp DNA fragment in nondenaturing 5% polyacrylamide gel after sonication at 44 kHz. Positions of labeled marker double-stranded DNA fragments of known sizes are shown with arrows on the electrophoretic patterns and on the sequence. The fragment was uploaded in gel before (lane 0) and after sonication (lanes 1–4) in 5 mM NaOAc (pH 7.0) for 2, 4, 8, and 16 min, respectively; (lanes 5–8) in 0.5 M NaOAc (pH 7.0) for 2, 4, 8, and 16 min, respectively; in 0.5 M NaOAc (pH 11.0) for 4 and 8 min, respectively; (lanes 9, 10), and in 0.5 M NaOAc (pH 5.0) for 4 and 8 min, respectively (lanes 11, 12).

The efficiency of cleavage slightly increased with ionic strength increasing from 5 mM to 0.5 M at pH 7.0. When pH was varied, cleavage was almost undetectable at pH 11.0, while its efficiency considerably increased at pH 5.0. This finding is explained by the fact that the double helix is partly unwound at alkaline pH, which increases the flexibility and the condensation of DNA. When sonication temperature was increased to 30°C, cleavage was almost completely suppressed (Figure 1). The fragment was cleaved to a significant extent at 30°C only at low pH.

To localize the DNA cleavage sites with single-nucleotide precision the sonication product was resolved by denaturing polyacrylamide gel. After sonication, the samples were combined with 180 mkL of solution containing 0.15 M NaCl, 50 mM Tris-HCl (pH 7.5), and 10 mM EDTA. The DNA was precipitated with ethanol, washed with 70% ethanol, dried, and dissolved in 1 mkL of 95% formamide. It was then heated for 1 min at 90°C, rapidly cooled down to 0°C, and applied on polyacrylamide gel containing 8 M urea. Electrophoresis was carried out for 55 min (100 W, 2500 V) at 60–70°C. Afterward, the gel was fixed in 10% acetic acid and dried on a glass plate pretreated with γ-methacrylpropyloxysilane. The dried gel was exposed to a luminescent screen and then scanned with a Cyclone Storage Phosphor System device (Packard BioScience). Cleavage pattern bands were assigned to particular nucleotide sequences of fragments by comparison with the lanes of "A+G" track DNA samples. Chemical cleavage of DNA with formic acid — diphenylamine reagent eliminates a purine from the cleavage site (Tate & Petersen 1975; Belikov & Wieslander, 1995). Thus, the bands seen in lanes "A+G" (on Figure 2 - lanes 1 and 18) correspond to oligonucleotides lacking the terminal purine. 3'-endlabeled fragments contain the uncharged 3'-OH group at the 3' end and the phosphate group, which carries two negative charges, at the 5' end. The bands observed after sonication of the DNA fragment exactly coincided with the bands observed after its chemical cleavage at purines (Figure 2). This result indicated that the products had phosphates at their 5' ends.

Figure 2 demonstrates the gel image obtained after sonication of the DNA fragments during various time periods. This image represents the results of ultrasonic irradiation of three fragments that differed in initial length (311, 251, and 218 base pairs, respectively) but shared the same base pair sequences. The left part of the gel contains lanes that correspond to cleavage of the longest fragment (lanes 1–6). The central part represents lanes corresponding to cleavage of the middle-sized fragment (lanes 7–12), and the lanes at the right side of the gel show the cleavage patterns of the shortest fragment (lanes 13–17).

Many double-strand breaks arose as early as within the first four minutes of sonication. Further sonication enhanced the cleavage pattern, but the ends of the fragment still remained intact. Increasing the irradiation time from 2 to 16 min leads to a sufficient increase in overall cleavage intensity for all three types of fragments. The addition of tiourea has no visible effect on the cleavage patterns (lanes 11 and 12). The same result was obtained when other free radical scavengers (dithiothreitol and sodium ascorbate) were added to the irradiated solution. On the other hand, adding 50% glycerol, which increased the viscosity of the solution by roughly 10-fold, led to a significant increase in cleavage intensities (lanes 3, 8, and 14). The cleavage patterns obtained by adding glycerol are similar to those obtained without it but with a longer irradiation time. Thus, increasing the viscosity of the solution leads to an overall increase in cleavage intensity but does not affect the relative intensities of cleavage. Figure 1 and Figure 2 show the positional effect (i.e., the attenuation of ultrasonic cleavage efficiency at sites that are closer to the ends of the DNA fragments. The significant role of the positional effect and the minor influence of free radical scavengers on the observed cleavage patterns have led us to a conclusion that the cleavage of DNA induced by free radicals in solution had been negligible in our experiments. The distinctive features of free radical cleavage of DNA on the gel (i.e., the emergence of overall cleavage background with no positional preference) were observed only at higher temperature.

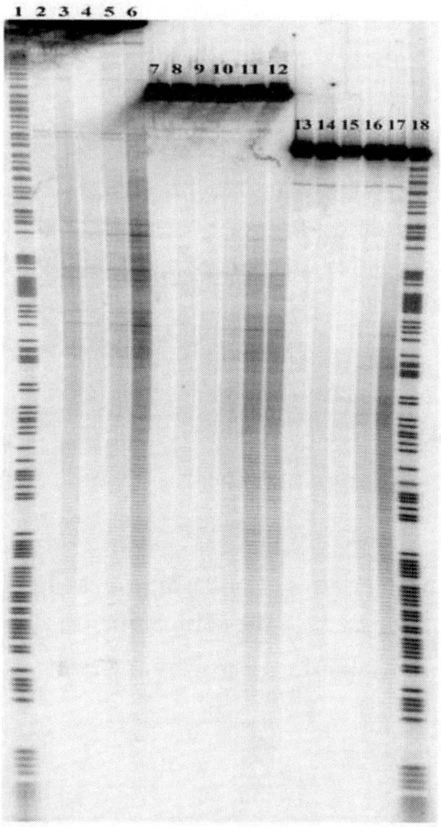

Figure 2. Cleavage pattern of DNA fragments in 6% - denaturing polyacrylamide gel after irradiation with ultrasound (22 kHz). Lanes: 1, 18 - chemical cleavage by purines; 2, 7, 13 - sonication of fragments for 2 min; 3, 8, 14 - sonication of fragments for 2 min in the presence of 50% glycerol; 4, 9, 15 - sonication of fragments for 4 min; 5, 10, 16 - sonication of fragments for 8 min, 6, 11, 17 - sonication of fragments for 16 min and 12 - sonication of fragments for 16 min in the presence of 0.5M tiourea.

2. GEL DATA ANALYSIS

The first stage of gel analysis involves the calculation of band intensities and their correlation with corresponding nucleotide sequence. To analyze gels we used SAFA package (Das et al., 2005) which allows aligning gel's lanes, calculating the overall intensity of each band and finding the correlation between the band sequence and corresponding nucleotide sequence. Band intensities are sensitive to various parameters of the experiment. Ultrasonic cleavage patterns demonstrate pronounced positional effect, i.e. the dependence of the band intensity value on its position in the gel. In order to analyze the sequence dependence of cleavage it is important to eliminate this general trend and operate with normalized values of cleavage intensities (i.e. efficiencies). This procedure might be performed using several approaches. We have compared the efficiency of several methods listed below: a) no trend elimination and normalization of band intensity by dividing band intensity by the mean value of bands intensities calculated using all analyzed bands of the lane; b) using the moving average method with various window sizes; c) describing the positional effect in terms of asymmetric gaussian functions; d) approximation of the trend with various degree polynomials (Nechipurenko et al., 2009). As a result of the comparison of these approaches performed during the analysis of DNA ultrasonic cleavage it was shown that the moving average method yields the best results.

The absolute value of individual band's intensity is further called "cleavage intensity", *I*, while the normalized value, *R*, and called "relative intensity of cleavage" or "cleavage rate". The normalization of band intensities was made by dividing their values by the local "basic band" intensity values, which were determined by the "moving average" method separately for each band. Namely, the intensity of each band was divided by the mean value calculated using the intensities of a number of adjacent bands. The optimal number of adjacent bands used for mean intensity calculation was found to be 31. Lowering this number results in increase of data points scattering while increasing the number of adjacent bands does not change the ratio of obtained relative intensities of cleavage, but leads to decrease in the number of analyzed data points.

Table 1. The sample characteristics of ultrasonic cleavage rates for dinucleotides. Designations: *N* – the sample size; $\overline{R}$ - the sample mean; *S* - standard deviation; $S_{\overline{R}}$ - standard error of mean

Type of dinucleotide	N	$\overline{R}$	S	$S_{\overline{R}}$	The 95% confidence limits	
					Lower limit	Upper limit
AA	1636	0.919	0.129	0.003	0.913	0.926
AC	1076	0.913	0.128	0.004	0.905	0.920
AG	1028	0.900	0.124	0.004	0.892	0.907
AT	1374	0.904	0.119	0.003	0.898	0.910
CA	1265	1.160	0.209	0.006	1.149	1.172
CC	1141	1.007	0.144	0.004	0.999	1.015
CG	1230	1.444	0.334	0.010	1.426	1.463
CT	1077	1.130	0.198	0.006	1.118	1.142
GA	1153	0.970	0.133	0.004	0.962	0.978
GC	1317	0.954	0.146	0.004	0.947	0.962
GG	1168	0.922	0.145	0.004	0.914	0.931
GT	1101	0.952	0.126	0.004	0.944	0.959
TA	1065	0.973	0.120	0.004	0.966	0.980
TC	1173	0.912	0.131	0.004	0.904	0.919
TG	1305	0.979	0.126	0.003	0.972	0.986
TT	1672	0.932	0.127	0.003	0.926	0.938

We analyzed the cleavage patterns for set of radiolabeled DNA restriction fragments with lengths from one hundred to several thousand base pairs (Grokhovsky et al., 2011).

To analyze the dependence between the nucleotide sequence and relative intensity of ultrasonic cleavage (*R*) of central phosphodiester bond in all possible di- and tetra-nucleotides we used ANOVA and nonparametrical methods (Kruskal-Wallis-test, Brown-Mood-test (Zar, 1999)) and the multiple comparison methods (T′-test, Tukey-Kramer-test, Dunn-test (Zar, 1999)). We noted that the nonparametric analysis yielded the same results as the parametric analysis.

The results of statistical treatment of the 20588 relative cleavage intensities for each of 16 dinucleotides are shown in Table 1, while the intensities for 256 tetranucleotides are presented in our previous paper (Grokhovsky et al., 2011). The effect of the dinucleotide type on cleavage rate is found to be statistically significant ($p \ll 0.05$). Moreover, there is a significant difference between the cleavage rates at complementary dinucleotides. Therefore, cleavage of particular phosphodiester bond does not always result in the cleavage of opposite phosphodiester bond in the complementary strand.

Importantly, the cleavage rates at dinulceotides d(CpC), d(CpT), d(CpA) and d(CpG) are significantly different from each other and also different from the cleavage rates obtained for all other dinucleotides. Cleavage rate of dinucleotide d(CpG) is the highest among all 16 dinucleotides.

Analysis of dependence on ultrasonic cleavage rate of nucleotide type at the 3’-position in each four groups of dinucleotides allows us to conclude that this dependence is significant in each of these groups ($p \ll 0.05$).

3. Evidence of Mechanochemical Nature of Ultrasonic Cleavage

The ultrasonic cleavage of DNA reported here is most likely the result of hydrodynamic shearing stresses caused by the collapse of cavitation bubbles (Basedow & Ebert, 1977, Suslick & Price, 1999). Their collapse results in a drastic increase of local temperature and pressure (Margulis, 1984; Didenko et al., 1999, McNamara et al., 1999). Shearing forces that act on the DNA fragments are thought to originate from high-velocity gradients of water near the collapsing bubble. High-velocity gradients in the streaming solution may cause mechanical deformation of the molecule by friction forces. Thus, the observed cleavage of DNA fragments most likely represents a complex mechanochemical process, which includes mechanical deformation of the molecule before the actual chemical reaction takes place (Basedow & Ebert, 1977). Because cavitational flows are accompanied by turbulence, any mathematical treatment of the problem is restricted to highly simplified models.

Figure 2 demonstrates that increasing the irradiation time leads to a sufficient increase in overall cleavage intensity of restriction DNA fragments, while the addition of free radical scavengers has no visible effect on cleavage patterns. (lanes 11 and 12). The same result was obtained when free radical scavengers (dithiothreitol and sodium ascorbate) were added to the irradiated solution. On the other hand, adding 50% glycerol, which increased the viscosity of the solution by roughly 10-fold, led to a significant increase in cleavage intensities (lanes 3, 8, and 14). The cleavage patterns obtained by adding glycerol are similar to those obtained without it but with a longer irradiation time. Thus, increasing the viscosity of the solution leads to an overall increase in cleavage intensity but does not affect the relative intensities of cleavage. This dependence of the ultrasonic cleavage intensity on the solution viscosity is one of the distinctive features of a mechanochemical reaction.

Figure 1 and Figure 2 also demonstrate the positional effect (i.e., the damping of ultrasonic cleavage) at sites that are closer to the ends of the DNA fragments. This relevant feature of ultrasonic cleavage patterns of DNA also supports the idea of mechanochemical nature of the cleavage process observed in our experiments.

4. Modeling Ultrasonic Cleavage of DNA

Evaluation of Cavitational Flow Gradients near Collapsing Bubble

Experimental data and calculations indicate that the collapse of cavitation bubble results in high gradient flows of the surrounding liquid. These high gradient flows are assumed to produce shear forces sufficient for polymer chain cleavage (Basedow et al., 1977). Today it is generally accepted that ultrasonic cleavage of polymers might be described with the model proposed by Thomas (Thomas, 1959). According to this model, the collapse of cavitation bubble leads to formation of radial velocity gradient which results in tension of the polymer molecules and their consequent cleavage.

In order to assess the values of radial velocity gradients near cavitation bubble we performed calculations of the collapse dynamics. The primary intent of this modeling was to obtain the lower estimates of the velocity gradient values which might originate during ultrasonic irradiation of DNA solutions in experiments described above.

The dynamics of cavitation bubble was modeled with Rayleigh-Plesset equation (Plesset et al., 1977):

$$R\frac{d^2R}{dt^2}+\frac{3}{2}\left(\frac{dR}{dt}\right)^2=\frac{1}{\rho}\left[P_i-P_\infty-\frac{2\sigma}{R}-\frac{4\mu}{R}\left(\frac{dR}{dt}\right)\right], \qquad (1)$$

where R – is the radius of the bubble, t – time, ρ – solution density, σ – surface tension of the liquid, μ – viscosity of solution, P_i – gas pressure inside the bubble, and P_∞ - pressure of surrounding liquid. The parameter values were estimated from experimental data corresponding to water temperature of 0-5 ^{0}C (σ= 75 10^{-3} N/m, μ=1.7 mPa ·s, $\rho=10^3$ kg/m^3).

The time-dependence of surrounding liquid pressure was described with harmonic function:

$$P_\infty = P_0 - P_A\cos(2\pi f t),$$

where $P_0=10^5$ Pa – is the atmosphere pressure, P_A =2.4 10^5 Pa– is the amplitude of ultrasonic wave pressure, f =22 kHz – frequency of ultrasound . P_A was obtained using equation $P_A=\sqrt{2I\rho c}$, here I – is the intensity of ultrasound which was estimated as 2 W/ cm^2, and c =

1480 m/s – is the speed of sound in water. Gas pressure inside the bubble was described using adiabatic approximation:

$$P_i = P_{i0}\left(\frac{R_0}{R}\right)^{3\gamma},$$

where P_{i0} – is the saturated vapor's pressure when the bubble reaches its maximum radius $R_0 \approx 80$ μm, $\gamma = 1.3$ is experimental value of adiabatic index for water vapor.

The pressure of saturated water vapor depends on the water temperature, although the gas pressure inside the bubble becomes the dominant factor of the bubble dynamics only at the final stage of the collapse. For reasons described below the final stage of the collapse was out of scope of our modeling. Adiabatic approximation allows one to obtain the higher estimate for gas pressure inside the bubble. It should be noted that vapor pressure is the only active factor which counteracts bubble collapse. It was shown (Flynn, 1964) that including heat and mass conduction equations in this model results in higher values of bubble's interface velocity and, thus, in higher values of radial velocity gradient during collapse.

Equation (1) was solved using the Runge–Kutta fourth-order method with the temporal step of $\Delta t = 10^{-10}$ s. The results of calculation were used for evaluating the radial velocity gradient in the surrounding fluid. Using the incompressible fluid approximation one can obtain the following equation for velocity gradient:

$$G \equiv \left|\frac{\partial V}{\partial r}\right| = 2\frac{R^2}{r^3}\left|\frac{dR}{dt}\right|, \qquad (2)$$

where G – is the absolte value of radial velocity gradient of fluid flow, V – radial velocity of flow at the distance r from the bubble's center ($r > R$).

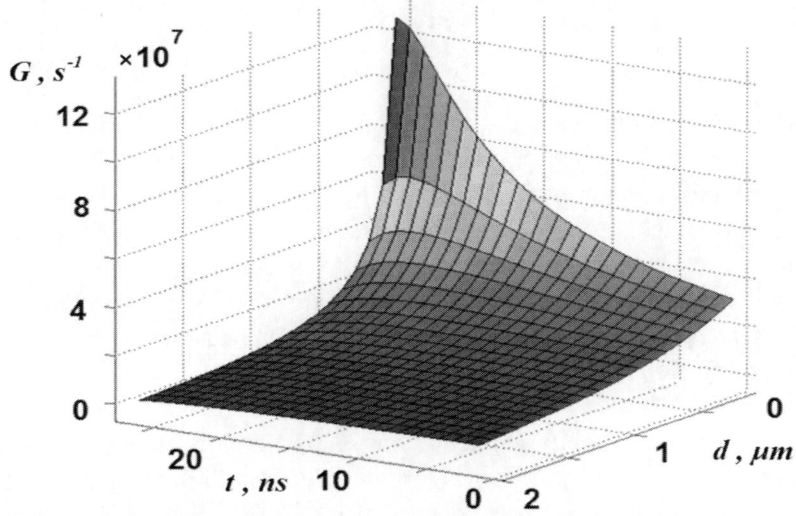

Figure 3. The dependence of radial velocity gradient G on distance from the bubble's interface d and time t.

Figure 3 shows the calculated dependence of radial velocity gradient of the flow near collapsing bubble. The value of gradient is higher for regions closer to the bubble interface and increases with time.

The results presented here correspond to particular stage of bubble collapse when its interface velocity values are less than the speed of sound in liquid. Hereafter we will refer to this stage as intermediate stage of bubble collapse. Processes which take place at the final stage of bubble collapse are out of scope of this model. These processes have nanoseconds characteristic times and result in catastrophic rise of temperature and pressure inside the cavitation bubble. This final stage of collapse involves chemical and optical effects of cavitation, formation of shock waves and microjets originating from instability of bubble shape at this final stage of collapse (Brennen, 1995).

Modeling DNA Dynamics in High Gradient Flow of Liquid

In order to describe the dynamics of DNA fragment in high gradient flow we used the beads on the strings model (Doi et al., 1986). DNA fragment was represented as the system of beads each having hydrodynamics radius R_D =1 nm connected with springs possessing rigidity constant k=6000 pN.

The frictional force acting on each bead from the liquid flow was described using the Stocks law:

$$F_s=6\pi\mu R_D V, \tag{3}$$

where F_s is the absolute value of drag force, R_D - hydrodynamic radius of the bead, V – the absolute value of bead velocity relative to the liquid, μ – dynamic viscosity of liquid.

Hence, the equation of motion for i-th bead of the system might be written as:

$$m\ddot{x}_i = \frac{k(x_{i+1}-x_i-l_0)}{l_0} - \frac{k(x_i-x_{i-1}-l_0)}{l_0} + \alpha(V_a-\dot{x}), \tag{4}$$

where m – is the mass of i th bead, x_i – its coordinate along axis x, k – rigidity of spring, l_0 – length of relaxed spring, α – constant describing the frictional force ($\alpha=6\pi\mu R_D$), V_a – velocity of the bead along axis x which is directed along the system of beads. It is worth noting that this model implies that the system of beads is linear and parallel to the liquid flow.

The length l_0 of each relaxed spring was taken to be equal $2R_D$. Thus, elementary spring unit corresponds to DNA block comprising 6 base pairs. In this case the value of frictional constant α corresponds to experimental value obtained from analysis of DNA oligonucleotides diffusion (Fernandes et al., 2002).

The dynamics of the model was evaluated in the homogeneous liquid flow along the x-axis which had a uniform velocity gradient along x-axis. Numerical integration of equation system (4) was performed using Verlet algorithm. The equilibrium distribution of tension through the system was calculated as the dependence of $F_i=k\ (x_{i+1} - x_i - l_0)$ on spring number i.

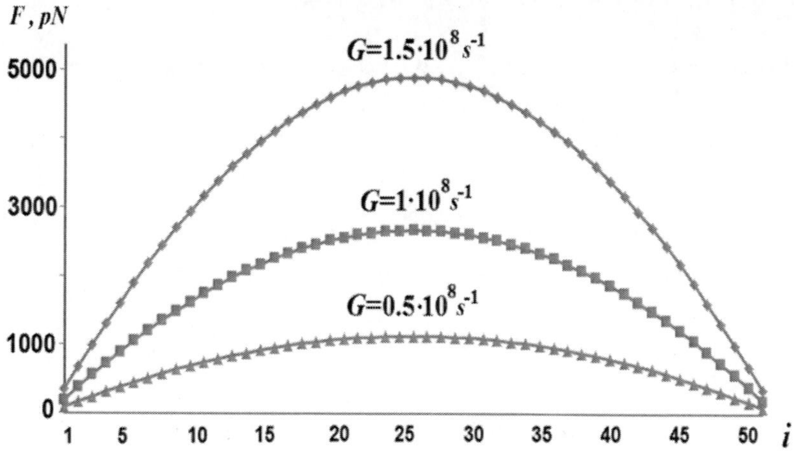

Figure 4. The dependence of tension force F of the spring on its number i. Tension profiles are superscibed with corresponding values of flow velocity gradients.

Figure 4 demonstrates the results of calculations obtained for three values of flow velocity gradients. The system consisted of 52 beads – an equivalent of approximately 300 base pair DNA fragment.

These results demonstrate that the presence of liquid velocity gradients exceeding 10^8 s^{-1} might produce tension force of several thousands of piconewtons. These values correspond to DNA ultimate stress limit values which range from 500 to 5000 pN (Smith et al., 1996). It should be noted that single-molecule studies of various polymers have shown that the rupture force of a single covalent bond is in the nanonewton range and depends logarithmically on the stretching rate (Bustamante et al., 2000).

The obtained values of tension force demonstrated strong dependence on such parameters as liquid viscosity μ, spring rigidity k and number of beads N. Liquid viscosity was chosen to correspond to viscosity of water at 0 °C (μ=1,7 mkPa·s) while the rigidity constant value k=6000 pN was obtained earlier using all-atomic molecular dynamics simulations (unpublished results). This value falls in experimentally obtained range (value of $k\approx1000$ pN was obtained at the single molecule DNA studies (Bustamante et al., 1994) and the value of $k\approx8000$ pN was estimated using the measurements of speed of sound in DNA fibers (Hakim et al., 1984)).

The obtained estimate for DNA tension of several thousands of piconewtons (in case of flow velocity gradients exceeding 10^8 s^{-1}) might result in significant conformational changes in DNA. Single molecule tension experiments demonstrate that conformational transition of DNA from B to so-called S-form occurs when the tension value exceeds 70 pN (Smith et al., 1996). However, those experiments were conducted on the time scales of seconds to minutes, while the timescales of DNA tension due to collapse-induced high gradient flows were about tenths of nanoseconds (see Figure 3). Conformational transition of DNA to overextended S-form which implies 1.5 fold increase of DNA contour length might be significantly hindered at such timescales due to hydration spine which tends to preserve DNA structure through the framework of hydrogen bonds between DNA and solvent molecules.

Modeling Mechanochemical Cleavage of DNA

Dissociation of chemical bond induced by external mechanical stress is a particular case of so-called mechanochemical reactions. Kinetics of mechanochemical cleavage is usually described using the transition state theory or empirical equations analogues to Arrhenius equation. According to modified form of transition state theory, the rate constant in case of action of catalyzing force *f* might be written as (Eyring, 1935 , Schmidt et al., 2008):

$$k_p = \frac{kT}{2\pi\hbar} e^{\frac{\Delta S}{k}} e^{-\frac{\Delta E_f}{kT}} , \qquad (5)$$

where k_p - is the reaction rate constant, *k*- Boltzmann constant, *T* – thermodynamic temperature, $\hbar$–Plank's constant, ΔS – entropy change upon transition state formation, ΔE_f- energy barrier for transition state formation in case of action of force *f* (modified activation energy):

$$\Delta E_f = E^* - E_0 - f.\Delta q , \qquad (6)$$

where E^* is the energy of transition state, E_0 – energy of ground state, *f* – external force component acting along the reaction coordinate (in case of dissociation or hydrolysis reaction this coordinate is assumed to correspond to the covalent bond length), Δq – reaction coordinate change required for formation of transition state.

In case of hydrolysis of DNA phosphodiester bond the value of activation energy is 26 kcal/mol, while the entropy change is about 34 cal/(K ·mol) (Schroeder et al., 2006). Thus, in the absence of tension the rate of hydrolysis reaction for C3'-O3' bond under normal conditions is about k_0=2 10^{-13} s^{-1}.

In case of mechanical tension *F* acting through the molecule, the rate of reaction according to (5) and (6) might be written as:

$$k_f = k_0 e^{\frac{f\Delta q}{kT}} \qquad (7),$$

where *f* – is the value of tension acting on the C3'-O3' bond of sugar phosphate backbone. This value depends on distribution of tension load *F* through molecular degrees of freedom. For example, this value depends on the angle between the vector characterizing overall tension *F* and direction of covalent bond. The value Δq stands for corresponding change of covalent bond length required for hydrolysis transition state to occur. Assuming that the major part of activation energy is attributed to proper deformation of the covalent bond C3'-O3', ·Δq might be estimated from the following equation:

$$V_m(.\Delta q) = \Delta E^* \qquad (8),$$

where $V_m(\Delta q)$ –is the Morse potential describing the energy shape of C3-O3' bond deformation: (Basedow et al., 1977): $V_m(\Delta q)=D\,(1-e^{-\beta\Delta q})^2$, where D=83 kcal/mol, β=2.58 Å^{-1}. According to (8), the value of $\Delta q \approx 0.3$ Å.

Using these equations it is possible to estimate the rate of DNA cleavage during the intermediate stage of cavitation bubble collapse. In order to perform such calculations we used the obtained temporal and spatial dependence of flow velocity gradient $G\,(r,t)$ and the results of calculations performed during modeling of DNA dynamics in uniform velocity gradient flow.

Consider that for moment of time t during bubble collapse the distribution of radial velocity gradient around the bubble is $G(r,t)$. Assuming that the concentration of DNA fragments is constant $n(r,t)=n_0$, the spherical layer with radius r and thickness dr will contain $n_0\,4\pi r^2 dr$ DNA fragments. Considering that a portion β of these fragments is aligned parallel to velocity flow, one can evaluate the tension F acting at the center of those fragments. Tension F depends on the value of local radial velocity gradient, i.e. $F(G(r,t))$. This overall tension distributes through various molecular degrees of freedom and is partially transmitted to every phosphodiester bond of DNA molecule. Thus, during time interval dt hydrolysis of particular phosphodiester bond will occur on average in dN number of DNA fragments: $dN=\beta\, n_0\, k_f\, 4\pi r^2 dr\; dt$, where k_f stands for the rate constant of mechanochemical dissociation reaction which is defined using Equation (7). The essential issue of this model is conversion of overall molecular tension F to its component f which corresponds to tension of particular phosphodiester bond:

$$f=\gamma F \qquad (9),$$

where γ – stands for transmission coefficient. This coefficient characterizes the coupling of overall tension with corresponding load exerted on particular bond. Since the tension in DNA molecule is distributed through both strands, $\gamma \leq 0.5$.

Thus, the portion k_{rel} of cleaved DNA molecules emerging during particular stage of bubble collapse might be written as:

$$k_{rel}=\frac{\iint\limits_{t,r} 4\pi r^2 n_0 \beta\, k_0 e^{\frac{\gamma F(G(r,t))\Delta q}{kT}}\, dtdr}{\int\limits_{r} 4\pi r^2 n_0 dr} \qquad (10)$$

where the integration is performed for defined interval of time. For every moment of time t the spatial integration is performed for the following range: $R(t)\leq r\leq R_{max}(t)$, where $R(t)$ – is the bubble radius at moment t, while $R_{max}(t) = R(t)+\Delta R$ represents the upper limit of spatial integration. Parameter ΔR was assigned a value of 50 μm since further integration does not produce any impact on k_{rel} value due to fast drop of velocity gradient values with increase of r (see Equation (2)).

To calculate the k_{rel} values we used the results of bubble collapse modeling. We considered the time interval of 80 ns which corresponds to intermediate stage of bubble collapse. The values $G(r,t)$ were calculated using Equation (2). The dependence $F(G)$

evaluated during modeling DNA dynamics in high gradient flow was approximated with 4th power polynomial using the least squares method. Integral (10) was calculated by direct summation using temporal step $\Delta t = 10^{-10}$ s for time variable and spatial step $\Delta r = 10^{-7}$ for variable *r*.

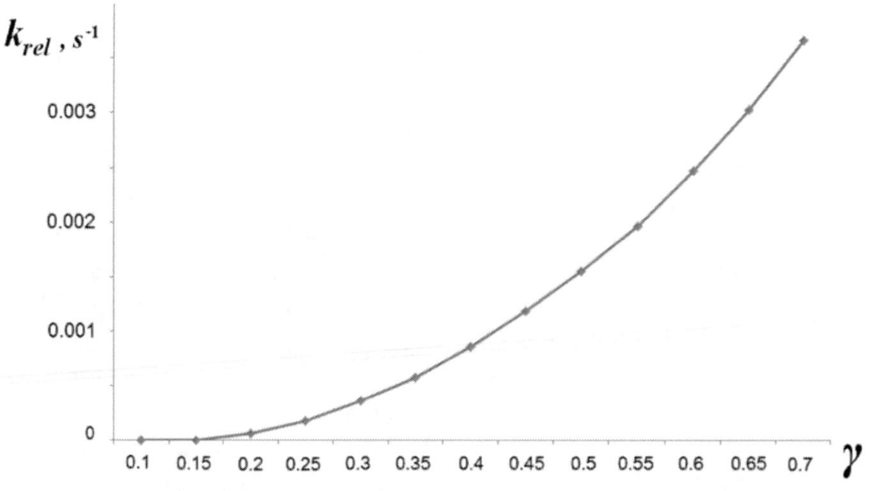

Figure 5. The dependence of DNA cleavage probability k_{rel} (recounted for 1 second) on the value of transmission coefficient γ.

The results of k_{rel} calculations for various values of transmission coefficient γ are presented below. The dependence of k_{rel} on γ given in Figure 5 was obtained with $\beta=1$. The k_{rel} values correspond to probability of DNA cleavage during 1 second, i.e. 22000 cycles of bubble collapse.

The ratio of DNA fragments cleaved during 1 second of ultrasonic irradiation was obtained from electrophoresis data and constituted 10^{-5}. Thus, theoretical values obtained for $\gamma=0.4$ corresponds to experimental data (if the probability of cavitation bubble formation during one cycle of ultrasound wave is about 10^{-2} or, equivalently, the concentration of cavitation nuclei in solution is about 10^3 ml^{-1}).

The results given at Figure 5 might be also compared to experimental data on ultrasonic cleavage of nicked DNA fragments, i.e. DNA molecules containing rupture of sugar phosphate backbone at particular site. It has been shown that ultrasonic cleavage of phosphodiester bonds opposite to initial breakage site increases by a factor of 10 to 30 compared to the ultrasonic cleavage of intact DNA fragments (Il'icheva et al., 2009). This factor is shown to depend on the nick position along DNA. In case of tension-induced cleavage the presence of rupture in one of DNA strands leads to rearrangement of tension distribution throughout the molecule: this stress becomes concentrated in intact strand of DNA. Such redistribution of the tension stress might be modeled as an increase in transmission coefficient: $\gamma_n = 2\ \gamma_0$, where γ_n and γ_0 – represent transmission coefficients in cases of nicked and intact DNA molecule, respectively.

The ratios of k_{rel} were determined using Equation (10). The calculated values of $k_{rel}(2\ \gamma_0)$ / $k_{rel}(\gamma_0)$ for $\gamma_0=0.2$, $\gamma_0=0.3$ and $\gamma_0=0.4$ were 16; 7 and 6, respectively. These results are consistent with experimental values ranging from 10 to 30.

The Equation (10) might be further applied for calculation of cleavage intensity profile, i.e. the dependence of ultrasonic cleavage efficiency of a particular phosphodiester bond on the number of these bonds in DNA fragments. The results of calculations performed for beads-on-the-strings model of DNA described above demonstrate that the ratio of central spring tension to the stretching force acting in the *i*-th spring does not depend on the flow velocity gradient for a wide range of values (data not shown). Hence, the value of tension in the *i*-th spring of the system might be calculated using the following equation: $F_i = F_{max} \cdot \alpha(i)$, where F_{max} – is the tension of the central spring, while the values $\alpha(i) \leq 1$ were calculated for $G=10^8\ s^{-1}$.

Figure 6 plots the dependence of k_{rel} on spring number obtained using Equation (10) with $\gamma = 0.4$.

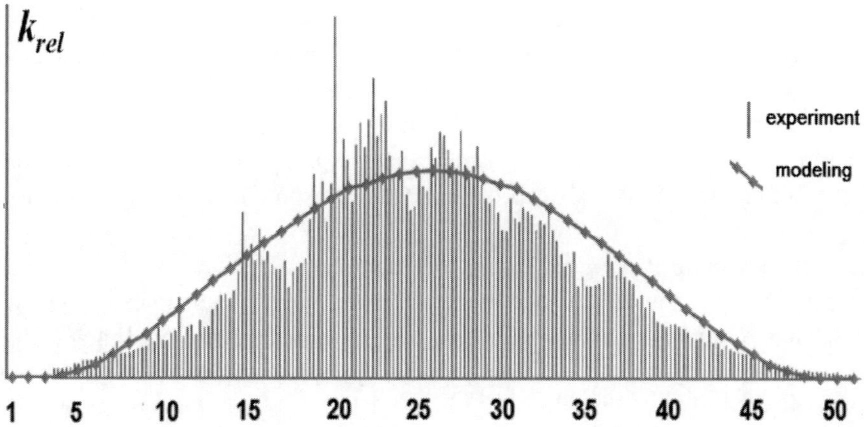

Figure 6. The comparison of theoretical cleavage profile with experimental data corresponding to lane 17 of Figure 2.

These results demostrate that this model of ultrasonic cleavage of DNA describes well the so-called positional effect which is evident from experimental cleavage pattern analysis.

Limitations of the Model

In the present work we have exploited a commonly used Thomas model addressing the ultrasonic cleavage of polymers. Our calculations demonstrate that in DNA cleavage may occur at the intermediate stage of cavitation bubble's collapse due to the formation of high velocity gradients in liquid flow. It should be noted that the cleavage of DNA under the effect of high-intensity ultrasound observed in our experiments might also be the result of shock waves generated during the collapse of cavitation bubbles. In case of shock wave–mediated cleavage, the mechanical deformation of molecules may be due to the drastic change of solvent pressure. We proposed the model of polymer cleavage under the action of cavitational shock waves (Gooberman, 1960) which describes the positional effect observed in experiments on ultrasonic degradation of polystyrene macromolecules. This model can be also applied to describe the ultrasonic cleavage of short DNA fragments (possessing length from tens to hundreds of nucleotide pairs) but this is a topic of a separate study.

According to the theory of mechanochemical reactions (Butyagin, 1971) there are two direct mechanisms for transition of mechanical energy to the energy required for chemical reaction to occur. The first mechanism implies the mechanical deformation of chemical bonds under the action of external force, while the second mechanism is associated with excitation of various vibrational modes during local molecular heat-up due to thermal relaxation of mechanical energy.

The observed features of ultrasonic DNA cleavage - the positional effect and selective cleavage of phosphodiester bonds, to all appearance, indicate that the first mechanism prevails in the described experiments. The breakage of DNA molecules can also occur due to extremely high and fast-acting forces leading to direct dissociation of phosphodiester bonds. Such forces may arise at the final stages of collapse of cavitation bubble, as well as by the shock waves generated during this collapse.

The results obtained with our model are qualitative in nature because of a number of factors that seriously complicate a more accurate description of ultrasonic DNA cleavage. Below we highlight the most important limitations of our approach.

The proposed models are based on approximation of diluted DNA solution, while actual DNA concentration is 5 to 10 μg/ml. At the same time the presence of various components (e.g. Na^+) in the buffer solution can lead to changes in some parameters used in our calculations.

The intensity of ultrasonic wave used during calculations -- 2 W/cm^2 can be overstated, as it was derived by comparison with experimental data on the yield of hydrogen peroxide received at the higher temperatures. Nevertheless, the presented here modeling results are likely to reproduce the lower estimate of flow velocity gradients since a number of factors increasing the velocity of bubble collapse were ignored. Moreover, the final stage of collapse was not considered in the proposed model.

Since the persistent length of DNA molecule is about 150 base pairs, using a linear fragment approximation for DNA molecule of 300 base pairs length is not fully justified. Nevertheless, in case of high velocity gradients near cavitation bubbles, polymer tension probably leads to linearization of molecule. The radial structure of the cavitational flow should contribute to the alignment of the fragments along the flow velocity vector.

At the same time, the conformational state of DNA at the presence of considerable tensile force acting on the time scale of a few nanoseconds is still debatable. The constant value of rigidity modulus k = 6000 pN, is an assumption since DNA rigidity actually depends on the value of stretching force (Bustamante et al., 2000).

Tension force F significantly depends on both the viscosity of solution and hydrodynamic radius of the bead. The estimates made for these parameters were rather approximate. All of these values, including transmission coefficient γ, affect the value of tensile force f stretching the covalent bond and are thus crucial for kinetics of cleavage process. Thus, changing the value of γ one can simulate the impact of such parameters as liquid viscosity μ, hydrodynamics radius of the bead R_D, fragment length L_0 and its stiffness k on the rate of DNA cleavage.

It should also be noted that the assumption of DNA hydrolysis as the primary mechanism of polymer cleavage reaction catalyzed with tensile force may be supplemented by considering the direct dissociation of covalent bond. This process may occur due to extremely high tensile forces that are likely to originate in a small volume at the final stage of cavitation bubble collapse, as well as during the propagation of shock waves. It should be also emphasized that in case of hydrolysis reaction the accessibility of phosphodiester bond for solvent molecules might be a substantial factor determining the rate of reaction. The solvent-accessible *surface area* of phosphodiester group affects the entropy factor in the expression describing hydrolysis reaction rate (5) and may depend on local conformational and dynamical properties of the sugar phosphate backbone.

The comparison of computation results with experimental data on overall DNA cleavage efficiency performed above is rather conditional, since the concentration of cavitation nuclei in the described experiments is unknown. This parameter can take values up to hundreds of thousands of bubbles per ml of solution, and substantially depends on its physicochemical properties. The derived estimate of bubble concentration of 10^3 ml^{-1} also depends on the value of transmission coefficient γ. If $\gamma = 0.2$, the cavitation nuclei concentration is estimated to be about 10^4 ml^{-1}. Thus, theoretical results imply fairly high values of cavitation bubble concentrations in order to correspond to experimental estimates of cleavage efficiency. It is important to note that the presence of DNA molecules can lead to the increase in concentration of cavitation nuclei in solution (Lentz et al., 2006).

Despite the noted limitations and drawbacks of developed model it qualitatively describes the distinctive features of observed ultrasonic cleavage of DNA: the positional effect of cleavage and the enhancement of overall cleavage for a) longer DNA molecules, b) lower solution temperature and c) higher viscosity of solution. Thus, the developed model might be used as the basis for more accurate description of the cleavage process.

5. Physical Properties of the B-DNA and Ultrasonic Cleavage Rate

The second feature of the obtained ultrasonic cleavage pattern is their sequence specificity. It cannot be explained by differences in stacking energies in the approximation of nearest-neighbor model, nor by the formation of unusual structures, for example kinks (Grokhovsky et al., 2011). Further we analyze the relationship between the obtained ultrasonic cleavage rates and known aspects of sequence-dependent B-DNA conformational dynamics. A number of conformational motions of sugar phosphate backbone are coupled with each other in a double-stranded DNA: the most relevant of them comprise S↔N interconversion of deoxyribose ring (Thomas et al., 1983 (1); Thomas et al., 1983 (2); Wartell et al., 1986; Rinkel et al., 1987) and BI↔BII motion (Fratini et al., 1982; Gorenstein 1987; Duranovich et al., 2004; Heddi et al., 2006). Sugar ring interconversion is transmitted to *trans*↔*gauche*$^+$ variations of δ torsion (see Figure 7) influencing the orientation of the C3’-O3’ bond which is involved in the cleavage process. This motion is also coupled with mutual rotation of the sugar around glycosidic bond (i.e. variation of χ) (Donohue et al., 1960) as well as with BI↔BII motion (Isaacs et al., 2001).

The enhancement of ultrasonic cleavage rate in phosphodiester bond following deoxycitidine is evident from the statistical analysis (Table 1). We conclude that some local property of the sugar phosphate junction of dinucleotides with deoxycitidine on the 5’-end d(CpN) distinguishes it from any other dinucleotides. This hypothesis is supported by some theoretical and experimental data. *Ab initio* quantum mechanical calculations (Foloppe et al., 1999 (1)) show the unique properties of deoxycytidine: the lowest potential energy region for glycosidic angle χ in deoxycitidine is found to be in the *anti*-region (relevant to the A-form DNA) not only for north, but also for south energy minima of deoxyribose. Any other nucleosides in the south energy minima prefer high-*anti* region for χ. Moreover, conformations of the north region for deoxyribose in deoxycitidine are characterized by slightly lower potential energy than in south region (Foloppe et al., 1999 (2)). Higher conformational entropy of the A-form DNA compared to the B-form for the GG/CC step distinguishes it from AA/TT step (Il'icheva et al., 2010), thus confirming the unique conformational behavior of the deoxycytidine. Hence, conformational preferences of the deoxycitidine are not in line with the regular conformational characteristics of the B-DNA (angle χ in high-*anti* region, pseudorotation in S-region). This contradiction may cause enhanced conformational S↔N motion in the deoxycitidine. Actually, this motion was registered by NMR ^{13}C spin relaxation study accompanied by molecular dynamics simulation (Duchardt et al., 2008).

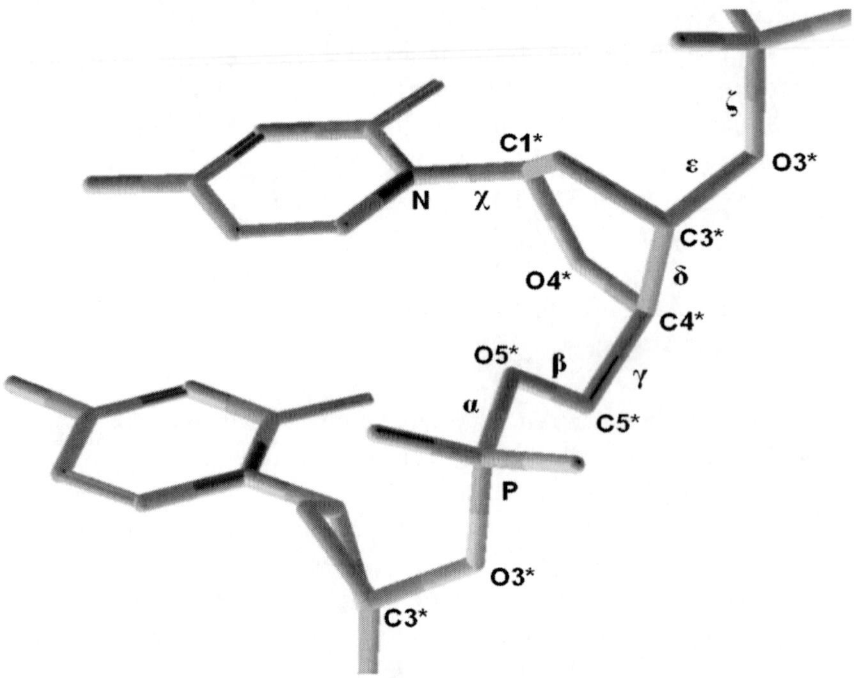

Figure 7. Schematic representation of dideoxynucleotide fragment with notation, which correspond to standard DNA structure description.

Another relevant conformational motion of the sugar phosphate backbone in B-DNA is determined by relation of ε/ζ torsions and comprise transitions between the canonical state BI (*trans/gauche*$^-$) and the minor state BII (*gauche*$^-$*/trans*). The preference in BI/BII conformation equilibrium is also sequence dependent (Heddi et al., 2010). The dynamic conformational coupling between equilibriums BI↔BII and S↔N in the B-DNA backbone was established by NMR (Isaacs et al., 2001). We analyze the reported statistics on BI/BII equilibrium ratios in different sequence context and receive evidence that there are no correlations between percentage of the minor BII state and the value of cleavage rate. Really, high percentage of the minor BII state in dinucleotides d(CpG) and d(CpA), which exhibit high cleavage rate in our experiments, is also found in d(TpG), d(GpG), d(GpC) and d(GpA), with appreciably lower cleavage rate. Moreover, cleavage rate in d(CpT) is high, but no cases of BII substate in d(CpT) have been reported, while d(ApG), which shows the minor cleavage rate, is characterized by high percentage of BII substate. Thus, BI↔BII equilibrium does not solely account for the observed enhancement of ultrasonic cleavage efficiency in d(CpN) dinucleotides.

The only peculiarity in conformational dynamics of the sugar phosphate backbone which is in line with ultrasonic cleavage rates is enhancement of S↔N interconversion of deoxyribose moiety in deoxycytidine. Hence, the conformational flexibility of deoxyribose is likely to be the dominant factor affecting ultrasonic cleavage rate of DNA.

The enhancement of cleavage might be the result of C3'–O3' bond reorientation during S↔N interconversion of deoxyribose: in the N-state this bond tends to align with local helical axis of DNA fragment (see Figure 8). When impulsive stretching force acts on the molecule, the decrease of angle between helical axis and particular C3'-O3' bond leads to increase in its tension f and thus lowers the activation energy of hydrolysis reaction (see Equation (7)). However it is necessary to further analyze in detail the second mechanism of ultrasonic influence, associated with excitation of various vibrational modes during local molecular heat-up due to thermal relaxation of mechanical energy.

Ultrasonic cleavage rate in d(CpC) dinucleotide is sufficiently lower comparing to another d(CpN) dinucleotides. This result might be due to the possible silencing of S↔N interconversion of deoxyribose in sequentially following deoxycytidines which is consistent with the NMR data (Isaacs et al., 2001). Such negative interference effect might be the result of reported positive correlation between BII conformation and N-type puckering of deoxyribose on the 5'-end of dinucleotide, and negative correlation of BII percentages in sequential phosphates. The assumption of the deoxyribose S↔N motion suppression in neighboring deoxicytidines might be also supported by the results of statistical analysis of relative intensities of the central phosphodiester bond in tetranucleotides d(N_1pCpN_3pN_4) (Grokhovsky et al., 2011). The results of multiple comparison tests (see Grokhovsky et al., 2011, Table S3) show that cleavage rates $\overline{R}$ for tetranucleotides of the type d(CpCpN_3pN_4) are minimal among the cleavage rates of d(N_1pCpN_3pN_4) tetranucleotides. Moreover, among 64 tetranucleotides of the type d(N_1pCpN_3pN_4) the lowest cleavage rate is found in d(CpCpCpC). Thus, if the intensity of S↔N motion is the dominant factor affecting the ultrasonic cleavage rate, then the negative interference effect of BI↔BII motion on the S↔N motion in sequentially following deoxycytidines is evident.

Ultrasonic cleavage rate seems to be modulated by sugar ring interconversion and on the other hand correlates with the local flexibility of B-DNA. On dinucleotide level of description ultrasonic cleavage rates are consistent with the range of experimentally obtained DNA deformability: PyPu > PuPu > PuPy (Okonogi et al., 2002). Summarizing cleavage rates $\overline{R}$ for dinucleotides referring to each of these three groups of steps (PyPu, PuPy and PuPy) (see Table 1) the following row of indices can be obtained: {2.32, 1.92, 1.86} which is consistent with the above mentioned deformability range. Thus, ultrasonic cleavage rate is likely to reflect sequence specific local flexibility of the B-DNA.

The cleavage rates for tetranucleotides are in line with some known data on the influencing of the context on conformational mobility in functional sequences of genome. For example, mobility of CpG is higher in d(TpCpGpA) than in d(ApCpGpT) (Lefebvre et al,

1995). The ultrasound cleavage rates for these tetranucleotides are 1.543 and 1.381, respectively. Tetranucleotide with a lower flexibility, d(ApCpGpT), is a part of the conserved gene promoter cAMP response element (CRE), recognized by a nuclear factor (CREB) that regulates the expression of many genes. Tetranucleotide with a higher flexibility, d(TpCpGpA), is a part of major mutation hot spots detected in the factor VI11 gene and responsible for hemophilia.

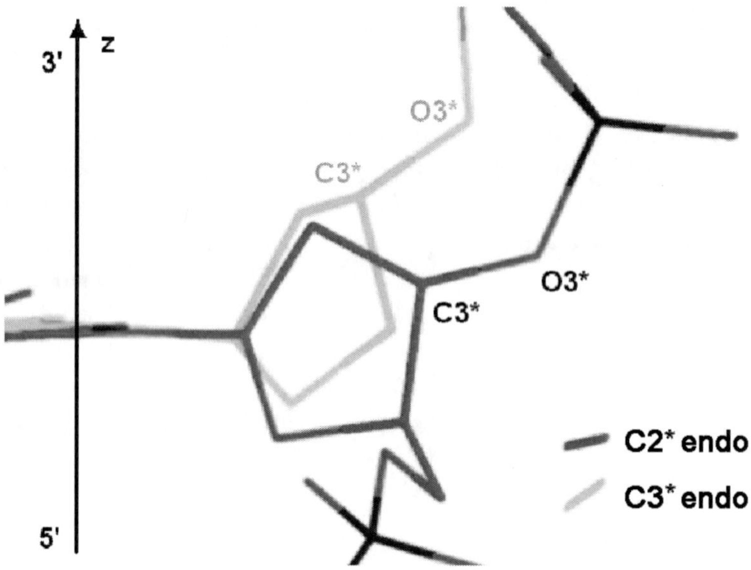

Figure 8. Schematic representation of C3'-O3' bond reorientation during S↔N interconversion of deoxyribose in B-DNA. The fragment of sugar-phosphate backbone with N-state of deoxyribose is shown faded while that with S-state is bright. The angle between DNA helical axis (shown as vertical z axis) and direction of C3'–O3' bond is lower in the N-state comparing to corresponding angle in the S-state. Structure NDB ID: BDL084.

CONCLUSION

The developed approach for studying ultrasonic cleavage of DNA allowed us to explain several effects. The positional effect was described using Thomas model of ultrasonic cleavage. Our calculations showed that intermediate stage of cavitation bubble collapse is capable of producing flow velocity gradients exceeding 10^8 s^{-1}. Interaction of polymers with flows possessing high velocity gradients might produce tension exceeding 10^3 pN which could be sufficient for DNA cleavage. The developed model seems to qualitatively reproduce the major effects of observed cleavage.

Cleavage rates, i.e. the mean values of the relative intensities of cleavage of the central phosphodiester bond in all 16 dinucleotides and all 256 tetranucleotides, were determined by multivariate statistical analysis. We observed a remarkable enhancement of cleavage rates of phosphodiester bonds after deoxycytidine, which diminished in the following row of dinucleotides: d(CpG) > d(CpA) > d(CpT) >> d(CpC). The cleavage rates for all pairs of complementary dinucleotides were significantly different from each other. The effect of flanking nucleotides in tetranucleotides on cleavage rates of all 16 types of central dinucleotides was also statistically significant. The sequence-dependent ultrasonic cleavage rates of dinucleotides were consistent with reported data on the intensity of the conformational motion of their 5'-deoxyribose. The sequence specificity of ultrasonic cleavage is the result of sequence-dependent conformational dynamics, and is likely modulated by the intensity of the sugar ring S<=>N interconversion. Sequence effects on conformational dynamics in any dinucleotide seem to propagate beyond mono and dinucleotide levels.

Hence, the relative efficiency of ultrasonic cleavage may serve as indicator of sequence specific flexibility in both strands of DNA. Each complementary chain can be characterized independently by the cleavage rate, and the diversity of conformational dynamics in both complementary chains can be estimated. Such numerical evaluation may be useful for identifying promoter regions in the genome (Nechipurenko et al., 2010) and assessing preferences for nucleosome positioning (Parker et al., 2009).

It is worth noting that the presence of islands with a high content of d(CpG) (possessing the highest value of ultrasonic cleavage rate) upstream and downstream of structural genes might provide a mechanism of gene shuffling during evolution. Epigenetic mechanisms based on d(CpG) methylation (Doerfler et al., 2006) might be the consequence of the unique properties of this step. The enzyme may recognize the unusual structure of the d(CpG) dinucleotide and flip cytosines out of the DNA helix during methylation more efficiently than the other bases.

It is worth noting that currently used methods for genome sequencing commonly use ultrasonic cleavage of the sample DNA and basically imply that this cleavage is non-specific. It is possible that the observed effect of sequence-dependence of DNA cleavage with ultrasound actually should be taken into account in order to avoid systematic errors during sequence assembly procedure.

GRANTS

This study was supported by the Program of the Presidium of the Russian Academy of Sciences on Molecular and Cell Biology and the Russian Foundation for Basic Research (projects 11-04-02001a).

REFERENCES

Basedow A. M., Ebert E. B. (1977) Ultrasonic degradation of polymers in solution. *Advances in polymers science*. Springer, Berlin/Heidelberg. 22: 83-148.

Belikov S., Wieslander L. (1995) Express protocol for generating G + A sequencing ladders. *Nuclei Acids Res*. 23: 310.

Brennen C. E. (1995) *Cavitation and bubble dynamics*. Oxford University Press.

Bustamante C. S., Smith B., Liphardt J., Smith D. (2000) Single-molecule studies of DNA mechanics. *Current Opinion in Structural Biology*. 10:279-285.

Butyagin P. Yu. (1971) Kinetics and nature of mechanochemical reactions. *Advances in chemistry.* (in Russian). 40:1935-1959.

Das R., Laederach A., Pearlman S. M., Herschlag D., Altman R. B. (2005) SAFA: Semi-automated footprinting analysis software for high-throughput quantification of nucleic acid footprinting experiments. *RNA* 11:344-354.

Davison P. F., Freifelder D. (1962) Studies on the sonic degradation of deoxyribonucleic acid. *Biophys. J.* 2:235–247.

Didenko, Y. T., McNamara W. B., Suslick K. S. (1999) Temperature of multibubble sonoluminescence in water. *J. Phys. Chem.* A103:10783-10788.

Doerfler W., Borm P. (eds) (2006) *DNA methylation: basics mechanisms*. Springer-Verlag, Berlin Heidelberg.

Doi M., and Edwards S. F. (1986) *The Theory of Polymer Dynamics*. Clarendon Press, Oxford.

Donohue J., and Trueblood K. N. (1960) Base pairing in DNA. *J. Mol. Biol.* 2:363-371.

Doty P., McGill B. B., Rice S. A. (1958) The properties of sonic fragments of deoxyribosenucleic acid. *Proc. Natl. Acad. Sci. USA*. 44:432-438.

Duchardt, E., Nilsson L., Schleucher J. (2008) Cytosine ribose flexibility in DNA: a combined NMR 13C spin relaxation and molecular dynamics simulation study. *Nucleic Acids Res*. 36:4211-4219.

Duranovich D., Hartmann B. (2004) DNA fine structure and dynamics in crystals and in solution: the impact of BI/BII backbone conformations. *Biopolymers.* 73:356-368.

Elsner H. I., Lindblad E. B. (1989) Ultrasonic degradation of DNA. *DNA*. 8:697-701.

Eyring H. (1935) *J. Chem. Phys.* 3, 107-115.

Fernandes M.X., Ortega A., López Martínez M.C., García de la Torre J. (2002) Calculation of hydrodynamic properties of small nucleic acids from their atomic structure. *Nucleic Acids Res.* 30:1782-1788.

Flynn H.G. (1964) Physics of acoustic cavitation in liquids. *Physical acoustics.* 1B:51.

Foloppe N., MacKerell A. D. Jr. (1999) Contribution of the phosphodiester backbone and glicosil linkage intrinsic torsional energetic to DNA structure and dynamics. *J. Phys. Chem.* 103:10955-10964.

Foloppe N., MacKerell A. D. Jr. (1999) Intrinsic conformational properties of deoxyribonucleosides: implicated role for cytosine in the equilibrium among the A, B, and Z forms of DNA. *Biophys. J.* 76:3206-3218.

Fratini A.V., Kopka, M. L. Drew H. R., Dickerson R. E. (1982) Reversible bending and helix geometry in a B-DNA dodecamer: $CGCGAATT^{Br}CGCG$. *J. Biol. Chem.* 257:14686-14707.

Gooberman G. (1960) Ultrasonic degradation of polystyrene. Part.1. A proposed mechanism for degradation. *J Polymer Sci*. .42:25-33.

Gorenstein, D. G. (1987) Stereoelectronic effects in biomolecules. *Chem. Rev*, 87:1047-1077.

Grokhovsky S. L. (2006) Specificity of DNA Cleavage by Ultrasound. *Molecular Biology* (Russian) 40:317-325.

Grokhovsky S. L., Il'icheva I. A., Nechipurenko D. Yu., Panchenko L. A., Polozov R. V., Nechipurenko Yu. D. (2008) Ultrasonic cleavage of DNA:quantitative analysis of sequence specificity. Biophysics (Russian) 53:250–251.

Grokhovsky S. L., Il'icheva I. A., Nechipurenko D. Yu., Golovkin M. V., Panchenko L. A., Polozov R. V., Nechipurenko Y. D. (2011). Sequence-specific ultrasonic cleavage of DNA. *Biophys. J.,* 100: 117-125.

Hakim H. B., Lindsay S. M., Powell J. (1984) The speed of sound in DNA. *Biopolymers.* 23:1185-1192.

Heddi B., Foloppe N., Bouchemal N., Hantz E., Hartmann B. (2006) Quantification of DNA BI/BII backbone states in solution. Implications for DNA overall structure and recognition. *J. Am. Chem. Soc.* 128:9170-9177.

Heddi B., Oguey C., Lavelle C., Foloppe N., Hartmann B. (2010) Intrinsic flexibility of B-DNA: the experimental TRX scale. *Nucleic Acids Res.* 38:1034-1047.

Il'icheva I. A., Nechipurenko D. Yu., Grokhovsky S. L.. (2009). Ultrasonic cleavage of nicked DNA. *J. Biomol. Struct. Dyn.* 27:391-397.

Il'icheva I. A., Vlasov P. K., Esipova N. G., Tumanyan V. G. (2010) The intramolecular impact to the sequence specificity of B→A transition: low energy conformational variations in AA/TT and GG/CC steps. *J. Biomol. Struct. Dynam.* 27: 677-694.

Isaacs R. J., Spielmann H. P. (2001) NMR evidence for mechanical coupling of phosphate BI-BII transitions with deoxyribose conformational exchange in DNA. *J. Mol. Biol.* 311:149-160.

Laland S. G., Overend W. G., Stacey M. (1952) Deoxypentose nucleic acids. Part III. Some effects of ultrasonic waves on deoxypentose nucleic acids. *J. Chem. Soc.* 303-310.

Lefebvre A., Mauffret O., Hartmann B., Lescot E., Fermandjian *S.* (1995) Structural behavior of the CpG step in two related oligonucleotides reflects its malleability in solution. *Biochemistry.* 34:12019–12028.

Lentz Y. K., Anchordoquy T. J., Lengsfeld C. S. (2006) DNA acts as a nucleation site for transient cavitation in the ultrasonic nebulizer. *J. Pharm Sci.* 95: 607-619.

Maniatis T., Fritsch E. F., Sambrook J. (1982) *Molecular Cloning: A Laboratory Manual.* Cold Spring Harbor, N.Y.: Cold Spring Harbor Lab. Press.

Margulis M. A. (1984) *Osnovi zvukohimii* (Russian). Moscow: Himiya.

McNamara W. B., Didenko Y. T., Suslick K. S. (1999) Sonoluminescence Temperatures During Multibubble Cavitation. *Nature.* 401:772-775.

Nechipurenko Yu. D., Golovkin M. V., Nechipurenko D. Yu., Il'icheva I. A., Panchenko L. A., Polozov R. V., Grokhovsky S. L. (2009). Characteristics of ultrasonic cleavage of DNA. *J. Struct. Chem. (Russ.),* 50: 1007-1013.

Nechipurenko D. Yu., Il'icheva I. A., Golovkin M. V., Panchenko L. A., Polozov R. V., Grokhovsky S. L. (2010) DNA conformational dynamics: approach to the physical mapping of genome. *Computer Research and Modeling (Russ.),* 2: 419–428.

Okonogi T. M., Alley S. C., Reese A. W., Hopkins P. B., Robinson B. H.. (2002) Sequence-dependent dynamics of duplex DNA: the applicability of a dinucleotide model. *Biophys. J.* 83:3446–3459.

Parker S. C. J., Hansen L., Abaan H. O., Tullius T. D., Margulies E. H.. (2009) Local DNA topography correlates with functional noncoding regions of the human genome. *Science.* 324:389-392.

Plesset M. S., Prosperetti A. (1977) Bubble dynamics and cavitation. *Ann. Rev. Fluid Mech.* 9: 145-185.

Pritchard N. J., Hughes D. E., Peacocke A. R. (1965) The ultrasonic degradation of biological macromolecules under conditions of stable cavitation. I. Theory, methods, and application to deoxyribonucleic acid. *Biopolymers.* 4:259-273.

Rinkel L. J., Altona C. (1987) Conformational analysis of the deoxyribofuranose ring in DNA by means of sums of proton-proton coupling constants: A graphical analysis. *J.Biomol. Struct. Dynam.* 4:621-649.

Schmidt S. W., Beyer M. K., Clausen-Schaumann H. (2008) Dynamic strength of the silicon−carbon bond observed over three decades of force-loading rates *J. Am. Chem. Soc.* 130: 3664–3668.

Schroeder G. K., Lad C., Wyman P., Williams N. H., Wolfenden R. (2006) The time required for water attack at the phosphorus atom of simple phosphodiesters and of DNA. *Proc Natl Acad Sci.* 11:4052-4055.

Smith S. B., Cui Y., Bustamante C. (1996) Overstretching B-DNA: the elastic response of individual double-stranded and single-stranded DNA molecules. *Science.* 271:795-799.

Suslick, K. S., G. J. Price. (1999) Applications of ultrasound to materials chemistry. *Ann. Rev. Mater. Sci.* 29:295-326.

Tate W. P., Petersen G. B. (1975). Stability of pyrimidine oligodeoxyribonucleotides released during degradation of deoxyribonucleic acid with formic acid — diphenylamine reagent. *Biochem J.,* 147: 439–445.

Thomas J.R. (1959). Sonic degradation of high polymers in solution. *J. Phys. Chem.* 63: 1725-1729.

Thomas G. A., Peticolas W. L. (1983) Flexibility of nucleic acid conformations. 1. Comparison of the intensities of the Raman-active backbone vibrations in double-helical nucleic acids and model double-helical dinucleotides crystals. *J. Am. Chem. Soc.* 105: 986-992.

Thomas G. A., Peticolas W. L. (1983) Fluctuations in nucleic acid conformations. 2. Raman spectroscopic evidence of varying ring pucker in A-T polynucleotides. *J. Am. Chem. Soc.* 105:993-996.

Wartell R. M., Harrell J. T. (1986) Characteristics and variations of B-type DNA conformations in solution: a quantitative analysis of Raman band intensities of eight DNAs. *Biochemistry* 25:2664-2671.

Zar J. H. (1999) *Biostatistical analysis.* Prentice Hall, New Jersey.

In: Ultrasonics: Theory, Techniques and Practical Applications ISBN: 978-1-62257-685-2
Editors: Hanako Ayabito and Mitsuko Katsukawa

Chapter 2

THE USE OF ULTRASONIC METHODS TO MONITOR THE EARLY AGE HYDRATION PROCESS OF CEMENT BASED MATERIALS

Gregor Trtnik[1,*] ***and Goran Turk***[2,†]
[1]IGMAT Building Materials Institute, Ljubljana, Slovenia
[2]University of Ljubljana, Faculty of Civil and Geodetic Engineering, Ljubljana, Slovenia

ABSTRACT

During the recent years comprehensive studies have been performed focusing on the possibility of using different ultrasonic techniques to monitor a hydration process and formation of structure of cement based materials at early ages. As a result, researchers all over the world have published many scientific articles dealing with this topic.

Among these, ultrasonic wave transmission and ultrasonic wave reflection methods are generally used. Using these techniques, the hydration process of various cementitious materials with different compositions can be measured continuously and non-destructively. Moreover, the methods have been proven to be very sensitive to the presence of different chemical and mineral admixtures in the material composition. On the basis of different ultrasonic parameters, procedures have been proposed to identify important milestones that appear during the formation of structure of the material, including initial and final setting time. Ultrasonic parameters have been successfully correlated to various material properties, namely the percolation of solid phases,

* Assit. Prof. Gregor Trtnik, PhD, works at IGMAT Building Materials Institute, Ljubljana, Slovenia. He is an assistant professor at University of Ljubljana, Faculty for Civil and Geodetic Engineering, Slovenia. He got his civil engineering degree in 2004 and earned a Ph.D. in 2009. His research interests are the possibility of using different US methods to monitor the setting and hardening process of concrete and other advanced techniques to study the hardening process of the materials. E-mail: grega.trtnik@igmat.eu.

† Prof. Goran Turk, PhD, got his civil engineering degree in 1987 and Ph.D. in 1994 at the University of Ljubljana, Faculty of Civil and Geodetic Engineering. He is currently a full professor at the University of Ljubljana. He teaches Statistics, Operational Research and Reliability of Structures. His research topics are: nonlinear computational mechanics, heat conduction and rheology of fresh concrete, fire analysis, statistical methods in engineering, reliability analysis and neural networks. E-mail: goran.turk@fgg.uni-lj.si.

compressive strength, elastic and shear modulus, degree of cement hydration, degree of capillary porosity, viscosity etc. Undoubtedly, these results improve practical application of the ultrasonic methods to a high degree.

In this chapter, an overview of some most important results dealing with the possibility of using ultrasonic wave transmission and ultrasonic wave reflection methods to monitor the setting and hardening process of cement based materials at early ages is presented.

HYDRATION PROCESS AND FORMATION OF STRUCTURE OF CEMENT BASED MATERIALS

During the last decade advances in measurement and computer technologies have allowed the development of various advanced methods to study the hardening process of different materials. Among these different ultrasonic methods turned out to be very effective in evaluating microstructural changes and are being used successfully also to monitor the hydration process and formation of structure of cement based materials (cement pastes, mortars, concretes).

Cement based materials are the most widely used building materials which change continuously their phase or state. After mixing main constituents (cement, water, aggregates, admixtures and additives), the material is viscous suspension. The state changes from fluid to solid material with increasing strength and stiffness. After few hours, cement based material converts to a solid material with high compressive strength.

These changes occur due to hydration process which is a general expression for the chemical binding of water in a mineral. As a result, a heat of hydration is evolved and different hydration products are formed, representing a solid phase in the microstructure of the material. The hydration process starts as a dissolution of cement particles with formation of new crystals. Generally, five stages can be identified during the hydration process (see Figure 1).

On contact with water cement grains immediately start to react. In this early *pre-induction period* C_3A is the most active phase. Cement particles are originally separated individually by water, and not connected to each other (Ye, 2003). Next, a period of a relatively low reactivity, known as a *dormant stage* occurs. The duration of this stage depends strongly on the curing (ambient) temperature, cement type (chemical composition), cement fineness, water/cement (w/c) ratio, and admixtures. At this stage, the cement paste remains plastic and is still highly workable. The time at the end of this stage is generally related to the *initial setting time* of cementitious material, which is considered as one of the most important characteristics of the materials at early ages. After the dormant stage, an *acceleration stage* begins and the reaction accelerates with time. In this period, the rate of C_3S hydration accelerates and a noticeable hydration of C_2S gets under way. This leads to the formation of different types of C-S-H gels. At a certain moment, the cement grains are connected by the C-S-H fibers and CH crystals (Ye, 2003) and the solid phases form a continuous solid body. This important milestone in the microstructure evolution is known as a threshold of solid phases percolation and can be considered as an indicator of the hardening state. Afterwards, the volume of the C-S-H and CH increases rapidly. Consequently, an intensive development of both total and connected solid phase can be observed. A final set generally occurs before the material shows a maximum rate of heat development at the end of the 3rd stage. The acceleration stage is followed by a *post-acceleration stage* and a *stage of a steady hydration*. In these periods the rate of heat release slows down. All hydration products are connected and the solid phase increase slowly. A schematic representation of the formation of structure of hardening cement paste is presented in Figure 2.

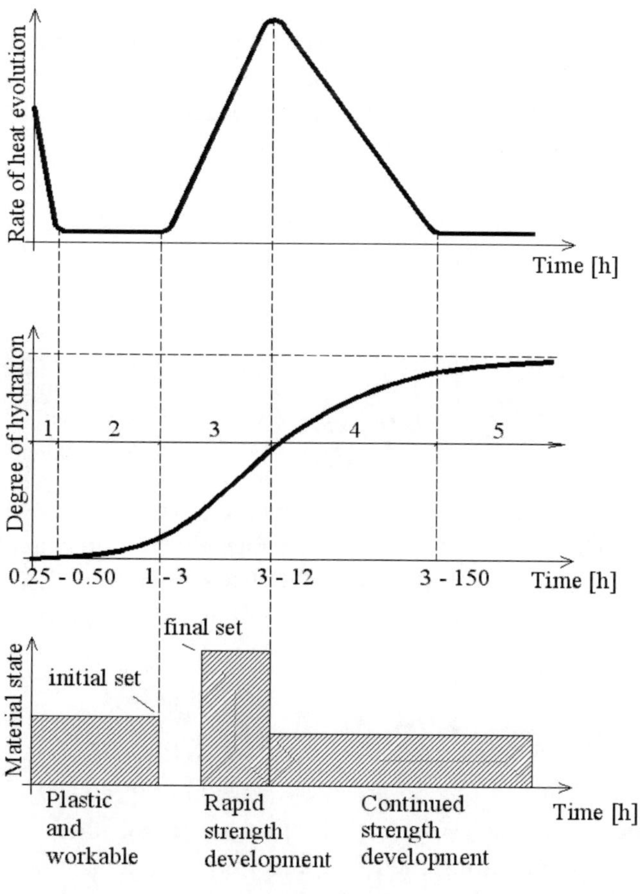

Figure 1. Stages during the hydration process (according to Schindler et al., 2002).

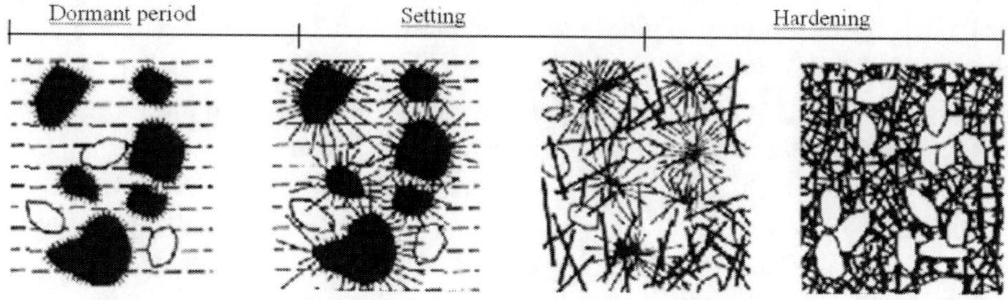

Figure 2. Formation of structure of hardening cement paste during the hydration process (according to Ye, 2003).

Different phases of the hydration process indicate that hydration is a complex chemical process involving the change from the viscous fluid state at the very beginning of the process to the solid state later on. Consequently, various material parameters, including shear modulus *G*, Young modulus *E*, and Poisson ratio change intensively during the early age hydration process. Significant changes in an ultrasonic wave propagation are therefore expected during the hydration process and formation of structure of cement based materials at early ages.

Lately, many studies have been performed with the objective to analyze the suitability of different ultrasonic methods to follow the hydration process and to monitor the development of microstructure of these materials at early ages. Among these, ultrasonic wave transmission method and ultrasonic wave reflection method are the most widely used and will be therefore described in more detail later in this section. Moreover, on the basis of the development of different ultrasonic parameters in time, many procedures have been presented to determine important milestones that appear during the setting and hardening of the material.

Ultrasonic Wave Transmission Method

Introduction

The principle of the ultrasonic wave transmission (USWT) method consists of a generating a pulse at one surface of the test specimen by an ultrasound transmitter. The according wave travels through the material and is picked up at the other side by a receiver. Generally, the incident and the transmitted pulses are recorded, converted to a digital signal and stored for further evaluation.

Experimental Set-ups

Various experimental set-ups using the USWT technique together with different moulds were developed to monitor the setting and hardening process of cement based materials.

At the Institute of Construction Materials (IWB), University of Stuttgart, an apparatus named FreshCon was developed aimed at investigating the setting and hardening process of cement based materials in quality assurance. Special software was also developed which shows the received ultrasound signal and their frequency spectrum using Fast Fourier Transformation (FFT) algorithm online during the experiment. As an addition, the changes in ultrasound velocity, energy, and the frequency content vs. material age are represented. With regard to the apparatus technology, the focal task has been also to develop an easily manageable container, which is made of Plexiglas. The container consists of two PMMA walls tied together by four screws with spacers. The mould is a U-shaped rubber foam element with the volume of approximately 30 cm^3. A detailed description of the FreshCon setup has already been presented by Reinhardt and Grosse (2004) and has been used by many research groups (Robeyst et al., 2008, Robeyst et al. 2009a, Robeyst et al. 2009b, De Belie et al. 2005). Similar U shape rubber has been used by Zhu et al. (2011). Ye et al. (2003, 2004) use a portable ultrasonic non-destructive digital indicating tester (Pundit Plus) with 54 kHz transducers. The transducers and temperature probe are integrated in a 150/150/200 mm^3 steel mould. In order to avoid the influence of the wall of the steel mould on the ultrasound propagation, two holes with a 54 mm diameter are made in the wall and the transducers are fixed by a PVC ring and coupled with the sample directly through a piece of plastic membrane glued on the inner wall of the mould. Four springs are used to adjust the transducer's contact pressure to guarantee good contact with the specimen. The temperature of the system is controlled by a cooling system with an accuracy of $\pm$ 0.1 K. A software controls the experiments and automatically records the hydration time and the ultrasonic pulse transition time at any predefined time intervals. This is of special importance in order to detect all the important phenomena that appear during the setting process of the material. Kamada et al. (2005) used a steel form with holes at the center of both sides. Ultrasonic sensors were secured on brass plates (50 mm long, 50 mm high, and 0.3 mm thick) with sensor holders to maintain contact pressure with the specimen. The ultrasonic pulse output from a multifunction synthesizer was amplified by using a bipolar power supply and then emitted with a voltage of 240 V and a pulse interval of 50 seconds. The ultrasonic sensors with a wide sensitivity range of 0-500 kHz and a resonant frequency of 140 kHz were used. Using this setup, wave velocity, maximum amplitude, and frequency distribution can be measured. Krauß and Hariri (2006) use an ultrasonic measuring device with the sensors integrated in the steel bucket. The transducers are protected by a plastic tube and held in place by a rubber seal ring which guarantees that no moisture is lost during testing. The transducers are coupled directly to the concrete and a commercial transition time measurement system was applied. Ultrasonic shear wave and compression wave transducers with a frequency of 50 kHz and 100 kHz were used, respectively. In the studies of Trtnik et al. (2008, 2009a) the Proceq CCT4 setup was chosen as a foundation which consists of a waveform generator board and two broadband transducers of central frequency of 54 kHz and 25 mm in diameter. The apparatus was adequately augmented to measure the v_P automatically at any desired intervals after casting. A special software was developed in order to collect v_P data with time. Cubic specimens of the size of 80/80/80 mm^3 were used and incorporated into a container of the extruded polystyrene foam Styrodur. Because of the high dissipation of the Styrodur materials, no waves propagating through the container wall were determined. The sensors were partly pushed into the specimens (about 5 mm) in order to ensure good contact between the transducers and the specimen and the acoustic coupling was enhanced by a layer of roller bearing grease. In recent studies performed by Trtnik et al. a new commercially available ultrasonic instrument is used. The received US signal, the trigger time and the corresponding P-wave velocity could be seen online during the experiment. Two broadband transducers of a central frequency of 150 kHz and 25 mm in diameter are used. Zhang et al. (2012) used an apparatus consisting of piezoelectric transducers, temperature controlling chamber, sample container, A/D converter and non-metal ultrasonic device. Aluminum foams were finally chosen to make the sample mould which combines the advantages of a metal with functional properties of a foam (excellent vibration and acoustic damping). The final container was designed into three parts including oil bath container for providing high temperature environment, a trapezoid shape sample mould with a volume of approximately 225 cm^3 for accommodating fresh cementitious materials, and ultrasonic transducers mounting device. Gabrijel (2010) used a commercially available Pundit instrument which consists of a waveform generator board and two broadband transducers of central frequency of 54 kHz and 50 mm in diameter. The apparatus was modified with the objective to measure the v_P

automatically and connected to the PC computer. A comprehensive work was performed aimed to develop a suitable container. At the end of his research, a container of the extruded polystyrene foam Styrodur was used with the dimensions of 150/150/70 mm^3 (for cement paste) and 150/150/100 mm^3 (for concrete).

The Use of USWT Method to Monitor the Setting and Hardening Process of Cement Based Materials

Typical Development of the v_p in Time

As indicated in previous section, different ultrasonic parameters can be measured continuously during the early age hydration process. However, the development of the velocity of ultrasonic longitudinal waves v_P in time ($v_P - t$ curve) is usually used to describe the formation of structure of various cement based materials. Based on the analysis of the experimental data, different researchers divide the $v_P - t$ curves into different characteristic stages. According to Ye (2003), Ye et al. (2003), Zhang et al. (2012), Kamada et al. (2005), Lee et al. (2004), and Chotard et al. (2001) three characteristic stages in the evolution of the v_P can be distinguished as indicated in Figure 3. However, on the basis of their derivatives, $v_P - t$ curves can be divided into four (Trtnik et al., 2008) or even five (Trtnik, 2009, Trtnik et al., 2009a) characteristic stages (Figure 3) which correspond very well with the classical definition of stages during the hydration process, determined on the basis of the heat release (see Figure 1). The first stage is generally referred to as the pre-induction and dormant period. During this period, ultrasonic waves rapidly propagate through the viscous suspension, resulting in a very low initial ultrasonic pulse velocity, which is often lower than in the air (340 m/s). This phenomenon was observed in many studies (Ye, 2003, Ye et al., 2003, Trtnik et al., 2008, Trtnik, et al., 2009a, Trtnik, 2009) and is caused by air entrapped in the material which causes a strong reflection and attenuates the sound wave velocity in a water/air system (Keating et al., 1989, Sayers and Dahlin, 1993). A detailed theoretical description of this phenomenon is given in (Robeyst et al., 2008). If the entrapped air is avoided, the velocity of the compression waves just after mixing is about 1500 m/s, representing a value close to the propagation velocity in water (Boumiz et al., 1996) which remains practically constant during the dormant period. However, in the case of a higher solid concentration (lower w/c ratio), slightly higher v_P was observed (Boumiz et al., 1996). During the second stage the first increase in v_P is observed which is not attributed to setting but to the formation of ettringite (Voigt, 2004). Although these needles do not create bonds between the cement particles they do fill pore space that was previously occupied by water with solid products, which is very beneficial for the propagation of P-waves (Voigt, 2004, Voigt et al., 2006).

According to Ye et al. (2003) the increase in v_P values during the 2nd stage can be attributed to a solid percolation path which is formed in the hydrating paste and caused the ultrasonic waves to propagate through the solid instead of liquid phase. Also, the entrapped air bubbles migrate to the surface of the specimen. Consequently, during this stage, it is believed that the v_P is controlled by the reduction of the amount of entrapped air bubbles and the amount of the connected solid phase, which can be indicated by the concept of bridge volume (Ye et al., 2003). Using HYMOSTRUC model (van Breugel, 1991, Lokhorst, 1999) it was determined that the correlation between the v_P and the bridge volume appears to be independent of the w/c ratio and curing temperature (Ye, 2003). During the third stage (acceleration stage), the solid phase becomes more and more connected, resulting into a rapid increase of v_P. At the end of this stage, an intensive development of the connected solid phase begins to slow down and stage four begins. In this stage, the influence of air bubbles drops and the v_P follows the total solid volume fraction (Ye, 2003).

During the final period (*period of a steady hydration*), most of hydration has been completed and the volume of pores has been filled up by the hydration products. Consequently, the v_P levels off and approaches its asymptotic value (Zhang et al., 2012).

Different stages in the v_P evolution in time indicate that the development of the v_P is closely related to the setting phenomena of cement based materials.

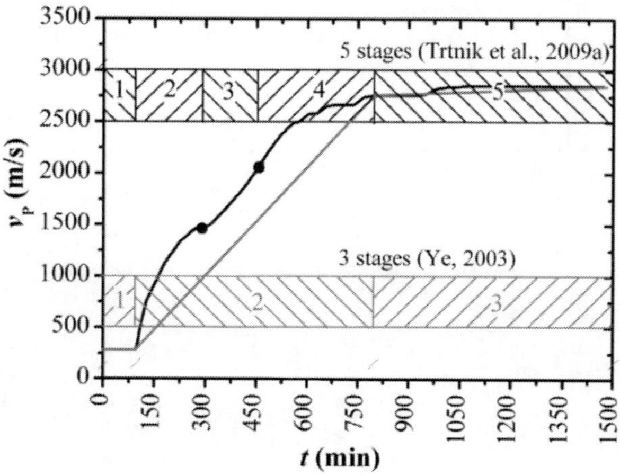

Figure 3. Definition of different stages on the $v_P - t$ curve (cement pastes). Solid dots represent inflection points.

Sensitivity of the v_P to the Material Composition and the Presence of Different Admixtures

Since the development of the v_P was found to be closely related to the setting phenomena of different cement based materials, a comprehensive research has been performed during the recent years with the objective to study the influence of different material composition and chemical and mineral admixtures on the evolution of the $v_P - t$ curves. It was found that $v_P - t$ curves are strongly affected by the w/c ratio (Trtnik, 2009, Trtnik et al., 2008, Trtnik et al., 2009a, Ye, 2003, Ye et al., 2003, Reinhardt and Grosse, 2004) and cement fineness (Trtnik, 2009, Trtnik et al., 2008, Trtnik et al., 2009a). Mixtures with a higher w/c ratio and lower cement fineness (*CF*) show lower v_P values. This can be attributed to higher volume of the total and the connected solid phase in the microstructure of the material in the case of lower w/c ratio and higher cement fineness (Trtnik et al., 2008, Ye, 2003) and consequently higher bridge volume (Ye, 2003; Ye et al., 2003). According to Reinhardt and Grosse (2004), the changes in the v_P are more evident in the case of a rapid hardening Portland cement. Higher v_P values were also observed in the case of a higher C_3A content (Trtnik, 2009, Trtnik et al., 2008, Trtnik et al., 2009a) and higher ambient temperature (T_A) (Trtnik et al., 2008, Trtnik et al., 2009a, Ye, 2003, Ye et al., 2003). Several researchers indicated that the v_P is also significantly affected by the presence of aggregates in the material, resulting in higher v_P in the case of larger aggregate and larger maximum aggregate size (Trtnik, 2009, Trtnik et al.,

2009b,). From Figure 4, influence of the cement paste composition on the development of the $v_P - t$ curves can be clearly observed.

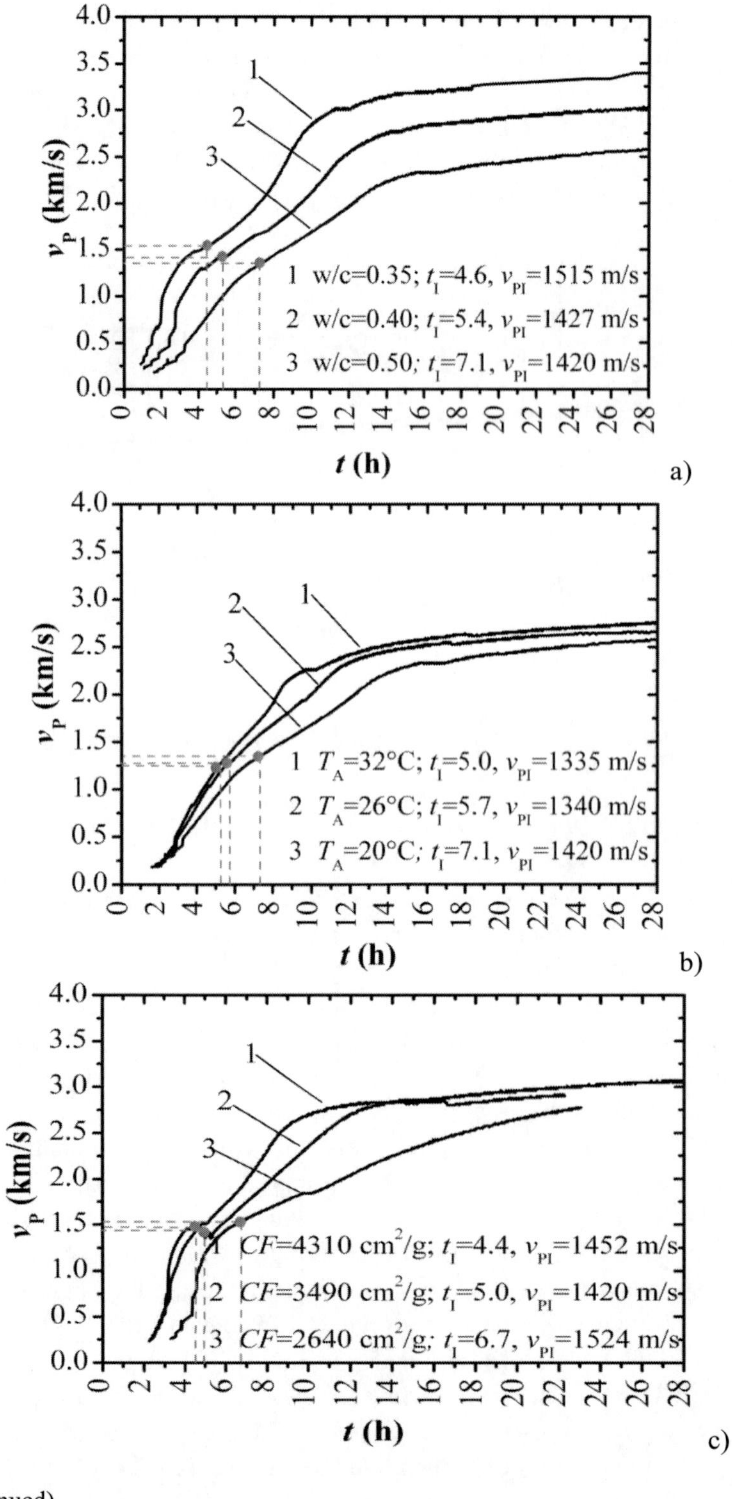

Figure 4. (Continued)

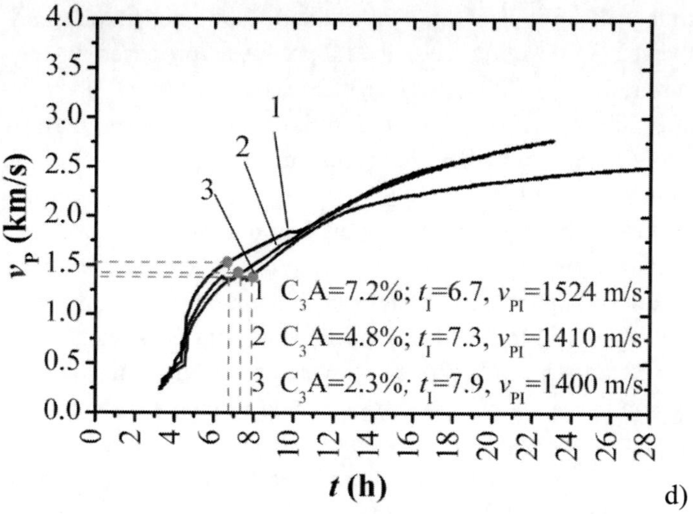

Figure 4. Development of the $v_P - t$ curves influenced by; a) w/c ratio, b) ambient temperature, c) cement fineness, d) amount of C_3A. Solid dots represent the 1st inflection point, t_I and v_{PI} stand for the initial setting time and corresponding v_P value, respectively.

Recently, some studies on the sensitivity of the USWT method to detect the presence of different admixtures have also been performed. Reinhardt and Grosse (2004) indicated that v_P – t curves were affected by the presence of an accelerator, a retarder, and an air entrained agent and concluded that the USWT method could be used very effectively when the production of admixtures should be continuously checked. According to their study, the effect of the retarder was the most clearer whereas the others were less prominent. Very recently, Zhu et al. (2011) investigated the effects of air voids on US P-wave propagation in early age cement pastes introduced by using air entraining admixtures. US velocity was strongly affected by air voids, resulting in lower v_P values. In the study of Zhang et al. (2012) effects of some mineral admixtures (fly ash, slag, and silica fume) on the evolution of the $v_P - t$ curves at early ages were analyzed. It turned out that silica fume addition increased the v_P velocity at the very beginning of the hydration process. A reverse phenomenon was observed when fly ash and slag were incorporated. Robeyst et al. (2008) studied effects of blast furnace slag on the development of the v_P and indicated that the v_P is generally lower if ordinary Portland cement is replaced by blast furnace slag or fly ash. De Belie et al. (2005) presented the possible use of USWT method to analyze the influence of some accelerating admixtures for shotcrete and indicated that US measurement were clearly sensitive to the effect of accelerator type and dosage of the setting behavior of mortar. A stepwise increase of the accelerator dosage resulted in increasing v_P values.

Determination of Different Milestones during the Formation of Structure of Hardening Cementitious Materials Using USWT Method

A number criteria have been suggested to indicate initial setting time based on the v_P measurements. A velocity range is usually given to determine the initial setting time since the velocity threshold values are practical and can be easily used during measurements. According to Lee et al. (2004), certain ranges of v_P can be used to estimate the initial setting times of mortars. In this study, 800−980 m/s and 920−1070 m/s velocity ranges were suggested to estimate the initial setting of ordinary Portland cement and fly ash mortars, respectively. However, some other velocity ranges were proposed in the case of concrete mixtures, for example 2300−2700 m/s (Herb, 2003) and 1000−1500 m/s (Van der Widen, 1990). Trtnik et al. (2008) concluded that the initial setting time of an arbitrary cement paste occurs approximately at the time when the v_P reaches the value closed to the P-wave velocity in water (i.e. 1450 m/s) – see Figure 4.

Similarly, various criteria have been proposed to determine the initial setting time on the basis of different characteristic points on the $v_P - t$ curves. The first inflection point on the $v_P - t$ graph was found to be an effective criterion to estimate the initial setting time of cement pastes (Trtnik et al., 2008, Trtnik, 2009) and mortars and concretes (Robeyst et al., 2008; Voigt et al., 2005). Kamada et al. (2005) showed that both initial and final setting occur within the 2nd stage on the $v_P - t$ curves, during which v_P increased monotonously. The initial setting time can be also defined by the time when the v_P starts to increase (concretes) (Reinhardt et al., 2000; Ye et al., 2003) or at the intersections of three straight lines tangent to the $v_P - t$ curve (Chotard et al., 2001).

The final setting can be also determined by velocity values or ranges. According to Reinhardt and Grosse (2004), Trtnik (2009), and Trtnik et al. (2009), the final setting time could be defined by an ultrasound velocity of 1500 m/s (concretes) and 1650 m/s (cement pastes), respectively. Lee et al. (2004), Herb (2003), and Van der Widen (1990) indicated that the final setting time can be determined by the time when the v_P reaches the values between 1200−1400 m/s (mortar), 2000−3000 m/s (concrete), and 2790−3180 m/s (concrete), respectively. Robeyst et al. (2008) related the final setting to the beginning of the third stage on the $v_P - t$ curves during which the v_P increases more gradually.

In the studies of Ye et al. (2003) it was shown that the development of the microstructure of cement based materials at early age can be described by ultrasonic pulse velocity measurements. For cement pastes, a unique correlation was found between the v_P evolution and the growth of a bridge volume, which is used as indicator for connectivity of solid phases (Ye, 2003). On the contrary, in the same bridge volume, lower v_P was found at higher curing temperatures for concrete mixtures (Ye et al., 2001). Both experimental and simulation results show that the time when the v_P starts to increase corresponds to the threshold of solid percolation, determined by Hymostruc program (Ye et al., 2004). Boumiz et al. (1996) indicated that at the time when the interconnected solid network spreads throughout the material, the shear wave may propagate throughout the materials. When all the solid phase was connected, the slow increase of the v_P followed the evolution of the total solid fraction.

Different studies have been also performed with the objective to determine some other milestones during the formation of structure of the material using USWT method. The v_P velocity was found to be correlated to a degree of cement hydration (Krauss and Hariri, 2006) and to the formation of different hydration products (Robeyst et al., 2011). Van der Winden (1990) used a number of practical criteria to determine the limits of workability of the material and suggested to define the limits of workability by the area where the v_P increased from 1000 m/s to 1500 m/s. Robeyst et al. (2008) indicated that within a time window of 30 min, the time when the penetration resistance starts to develop corresponds with the inflection point on the $v_P - t$ curves. According to Voigt (2004), especially at early ages, the pulse velocity increases at a much higher rate than the adiabatic heat release. A power law trend was found between the chemical shrinkage, the adiabatic heat release and the P-wave velocity, regardless of the addition of admixtures, the type of the test material or the w/c ratio.

For mixtures cured at different temperatures, almost linear relationship was found between v_P and compressive strength exceeding 10 N/mm^2 at early stage (Ye et al., 2001). However, at very early ages, observations of a rapid increase of v_P accompanied with very small changes in compressive strength of concrete were made (Voigt et al., 2006, Elvery and Ibrahim, 1976). This relationship was found to exhibits an exponential relationship at later ages (Trtnik, 2009, Trtnik et al., 2009b). Voigt et al. (2006) found out that measurements conducted with S- waves in transmission mode are in very close correlation to the evolution of the compressive strength of the tested mortar at very early ages.

Monitoring the Setting Process of Cement Based Materials Using the Entire Signal

Lately, some other US parameters were proven to be appropriate for monitoring the hydration process and formation of structure of cementitious materials. These parameters consider the entire received signal and could be therefore equally or better suited when compared to P-wave velocity (Robeyst et al., 2009). Moreover, they depend less on accurate determination of the signal onset time which is an advantage over the classical v_P measurements. Grosse and Reinhardt (2000) indicated that using Fast Fourier Transformation of the received signal, two dominant frequencies can be obtained during the hardening of the concrete. Similar findings were reached by Kamada et al. (2005) where cement pastes were studied. It was shown that at the very early age after mixing when the concrete is still a fluid-like suspension only low frequencies with relatively small amplitudes can be detected while the high frequencies are completely damped (Boumiz et al., 1996, Grosse and Reinhardt, 2000, Robeyst et al., 2009a). During the first hours no significant changes are noticed in the spectra which can be attributed to the dormant period of the hydration process. When material starts to set the high frequencies gradually appear. De Belie et al. (2005) indicated that the frequency shift is influenced by the presence of the accelerator type and dosage. In this study, the change of the received ultrasonic wave energy in time was also used as an indicator of the development of the setting process. The energy is proportional to the sum of the squared amplitudes of the received ultrasound signal. Similar study was performed by Robeyst et al. (2009b) who indicated that increase in the ultrasonic energy during the setting process of the material is retarded if ordinary Portland cement is replaced by a blast furnace slag or fly ash. For the materials containing high percentages of this cement replacing additives a second increase in the energy vs. time graph can be detected. They concluded that the wave energy transmission is a useful quantity to monitor the setting behavior of cementitious materials. The thresholds E/Eref = 0.02 and 0.13 were proposed to easily determine initial and final setting of the material, respectively. Using exploratory data analysis, the utility of the frequency spectra to monitor the setting of fresh concrete was also proven (Robeyst et al., 2009a). Very recently, Robeyst et al. (2011) performed a comprehensive research with the objective to establish a relation between different ultrasonic parameters (pulse velocity, frequency spectra and energy ratio) and fundamental changes in the cement paste. Multi-way analysis was performed for this purpose. Six parameters related to the microstructure development were simulated with an adjusted version of the pixel model CemHyd3D (Bentz, 1999), namely the solid percolation, pore space percolation, degree of cement hydration, degree of total binder hydration and formation of different hydration products. According to this study, the degree of cement hydration can be best correlated to the v_P, but the models based on the frequency spectra an energy ratio performed almost equally well. On the contrary, the solid percolation, the degree of total binder hydration, and the formation of type

I and type II hydration products were best predicted by the models based on the frequency spectra. A poor relationship was established between the pore space percolation and ultrasonic parameters which can be attributed to the fact that the effect of pore space depercolation on early-age ultrasonic measurements might be minor (Rapoport et al., 2000).

ULTRASONIC WAVE REFLECTION METHOD

Introduction

The first application of the ultrasonic wave reflection (USWR) method to the area of cementitious materials was reported by Stepišnik et al. (1981) and was extended further by Valič (2000). The method is based on measuring a shear wave reflection coefficient r^* of a pulsed ultrasonic shear wave formed at a buffer material and cement based material interface. Since cement based material changes its state from a fluid-like suspension to a solid material, r^* is changing in the process of hydration and therefore gives specific information on the hydration kinetics and formation of structure of the material. Reflection coefficient can be directly related to an acoustic impedance of the buffer and test material, which is calculated by:

$$r^* = \frac{Z_2 - Z_1}{Z_2 + Z_1}, \tag{1}$$

where Z_1 and Z_2 are acoustic impedances determined by the viscoelastic properties of each of the two interface-forming media. If a medium of known Z is used as the buffer material, the complex reflection coefficient defined by Eq. (1) can be written in the form (Valič, 2000):

$$r^* = r(\cos\varphi + i\sin\varphi) , \tag{2}$$

where r is the magnitude and φ the phase angle between the incident and the reflected waves. Since the relative phase changes during the hydration are generally quite small ($\varphi \approx 0$), the Eq. (2) can be simplified to

$$r^* = r \tag{3}$$

which makes the instrumentation less complex and difficult (Valič, 2000). Valič suggested that in presenting the r data in time t, it is better to plot the change of the reflection coefficient according to

$$\Delta r(t) = 1 - r(t) \tag{4}$$

which is increasing with the continuing hydration process and therefore resembles the diagrams of the increasing degree of hydration and other microstructural properties of cementitious materials (Valič, 2000).

Experimental Set-ups

A basic model of an apparatus using the pulse USWR method was first described by Valič (2000). Lately, several ideas for improvements surfaced with the objective to develop an instrument which can be easily and effectively used directly in-situ. As a result, a new apparatus USWR-4 Hardening meter was constructed in a multi-head version with four measuring heads operating simultaneously, allowing the measurements to be performed at four different locations on the concrete structure (Figure 5).

A detailed description of the principles of operation and performance of the apparatus can be found in (Valič, 2000). The USWR-4 Hardening meter consists of a main frame box with transmitter/receiver electronics, A/D converter board and power supply, measuring heads, and PC computer with suitable software. Measuring heads are of rugged construction and consist of a cylindrical aluminum body ($\Phi = 30$, $l = 40$ mm) in which a very pure fused quartz rod of rectangular cross-section ($a = 10$, $b = 16$ mm) and length $l = 50$ mm is rigidly fastened. On one end (bottom) a PZE ultrasound transducer, acting as a transmitter and receiver, is hard bonded. On the other end (top), with a measuring surface of 2 cm^2, the sample to be tested is smeared which can be easily performed at laboratory conditions. Alternatively, the quartz rods can be simply pushed into the sample or installed on the surface of the concrete element, allowing the measurements to be performed directly in-situ. USWR-4 Hardening meter was used in studies performed by Trtnik (2009) and Trtnik et al. (2009a).

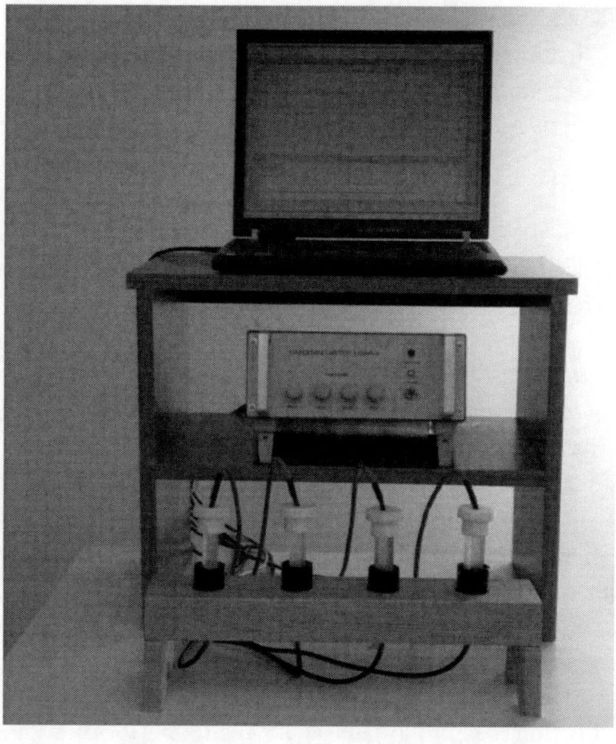

Figure 5. USWR-4 Hardening meter for measuring the shear wave reflection coefficient on the quartz/cement paste interface.

On the basis of this apparatus, various other experimental setups have been developed. Voigt et al. (2004, 2005, 2006), Sun (2005), and Sun et al. (2004) use a setup which consists of a laptop computer with a suitable software, a pulser/receiver unit, two transducers with 2.25 MHz central frequency, and a steel plate with a thickness of 12 mm as a buffer material. The unit excites the transducer and transmits the information of the received reflections from the transducer to the PC (Voigt, 2004). In other studies performed by Sun et al. (2006), a fused quartz plate with a thickness of 10 mm was used as a buffer which has a smaller acoustic impedance than steel, resulting in a larger bandwidth of the $\Delta r - t$ curves.

Ozturk et al. (2006) used a heat insulated test setup with a normal-incidence L-wave transducer type DEUTSCH S40HB0.1-0.3 with a center frequency of 200 kHz. The transducer is coupled on the acrylic glass plate with a thin layer of vaseline. A mould is glued on the acrylic glass plate and a water basin with constant 20°C water is placed on top of the specimen in order to achieve isothermal conditions.

Chotard et al. (2001) developed an instrument with the transducers which sends wideband pulses of longitudinal ultrasonic wave with a central frequency of 1 MHz. The dimensions of PMMA mould were 100/50/10 mm. The longitudinal wave velocity was measured in the direction at the shortest dimension. The PMMA mould and a broad-band ultrasonic transducer with 1 MHz central frequency was also used in the studies of Subramaniam et al. (2005).

Recently, Subramaniam and Wang (2010, 2011) presented an experimental system which included a function generator, an oscilloscope, a computer, and a test probe which comprises of ultrasonic transducers attached to a buffer plate made of fused quartz. The fused quartz plate was fabricated with angled faces at angles in the range from 0 to 60 degrees with respect to the horizontal. At each angle of incidence, a matched pair of direct contact, ultrasonic shear-wave transducers with a nominal center frequency of 1 MHz was mounted on the opposite angled surfaces of the buffer plate.

The Use of USWR Method to Monitor the Setting and Hardening Process of Cement Based Materials

Typical Development of the Δr (RL) in Time

Similar to the $v_P - t$ curve in the case of USWT method, the $\Delta r - t$ or $RL - t$ curves are usually used to describe the development of the hydration and microstructure process of different cementitious materials. Sun (2005) and Sun et al. (2004) indicated that typical $RL - t$ curve can be divided into three phases (see Figure 6). During the 1st phase the paste is considered as cement particles suspended in water. Almost no *RL* (reflection loss) can be noticed during this dormant stage since the reflection loss development is governed by the connectivity of the cement particles (Voigt et al., 2005). With an ongoing hydration, the amount of solid phases increases rapidly, resulting in an intensive increase of *RL* values during the 2nd stage. At a certain degree of hydration, the final stage begins. Since the development of the reflection loss at the later ages was found to follow that of the total solid phase, a moderate or low growth rate of the *RL* values are observed during this final stage (Voigt et al., 2005). As indicated by Sun (2005), two turning points can be observed on the $RL - t$ curve, even though they are not well defined. Analogically to the $v_P - t$ curves, five stages, indicating in Figure 6, can be defined on the $\Delta r - t$ curves if $\Delta r - t$ curves together with their derivatives are analyzed (Trtnik, 2009, Trtnik et al., 2009a).

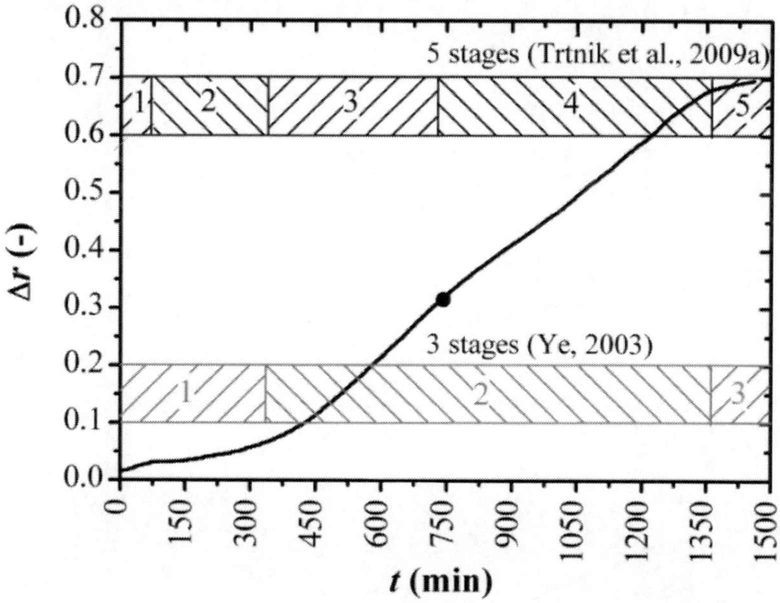

Figure 6. Definition of different stages of the Δr – t curve (cement pastes). Solid dot represents an inflection point.

Sensitivity of the Δr (RL) to the Material Composition and the Presence of Different Admixtures

The sensitivity of the USWR method to monitor the setting process of cement pastes was studied by many researchers. It was found that mixtures with a higher w/c ratio show lower values of Δr (*RL*) (Valič, 2000, Trtnik, 2009, Trtnik et al., 2009a, Sun, 2005). Further, higher curing temperatures resulted in a more rapid increase of Δr (*RL*) values at very early ages. A well known cross over effect was also observed (Trtnik et al., 2009a, Sun, 2005), which was previously indicated by other researchers in their studies of the influence of the curing temperature on the development of concrete compressive strength and adiabatic hydration curves (Van Breugel, 1991, Korhonen, 2002). Higher Δr values were observed in cases of higher C_3A content (Trtnik et al., 2009a, Trtnik, 2009) and higher cement fineness (Valič and Stepišnik, 1998, Trtnik et al., 2009a, Trtnik, 2009). In Figure 7, the development of the $\Delta r - t$ curves influenced by w/c ratio, ambient temperature, cement fineness, and amount of C_3A is presented. Valič (2000), Valič and Vuk (2000) and Sun (2005) indicated that $\Delta r - t$ and $RL - t$ curves are strongly affected by the presence of accelerator and retarder in the composition of the material. A sensitivity of the USWR method to the presence of different cement types (Valič and Stepišnik, 1998, Valič, 2000, Trtnik et al., 2009a), including aluminous and fast hardening Vezur cement pastes and mortars (Valič and Stepišnik, 1998) and mineral admixtures of calcite and blast furnace slag (Valič and Vuk, 2000), was also presented.

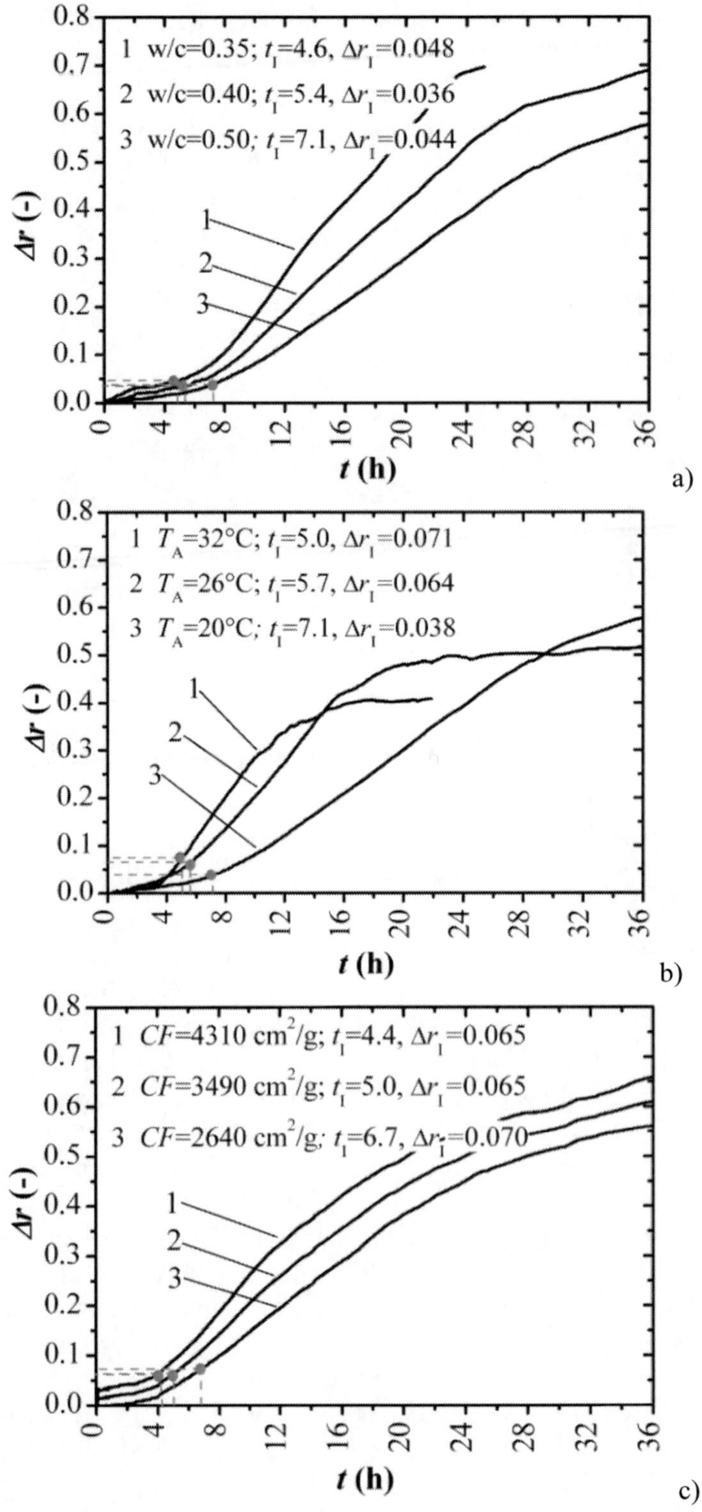
1 w/c=0.35; t_I=4.6, Δr_I=0.048
2 w/c=0.40; t_I=5.4, Δr_I=0.036
3 w/c=0.50; t_I=7.1, Δr_I=0.044
Δr (-)
t (h)
a)
1 T_A=32°C; t_I=5.0, Δr_I=0.071
2 T_A=26°C; t_I=5.7, Δr_I=0.064
3 T_A=20°C; t_I=7.1, Δr_I=0.038
Δr (-)
t (h)
b)
1 CF=4310 cm^2/g; t_I=4.4, Δr_I=0.065
2 CF=3490 cm^2/g; t_I=5.0, Δr_I=0.065
3 CF=2640 cm^2/g; t_I=6.7, Δr_I=0.070
Δr (-)
t (h)
c)

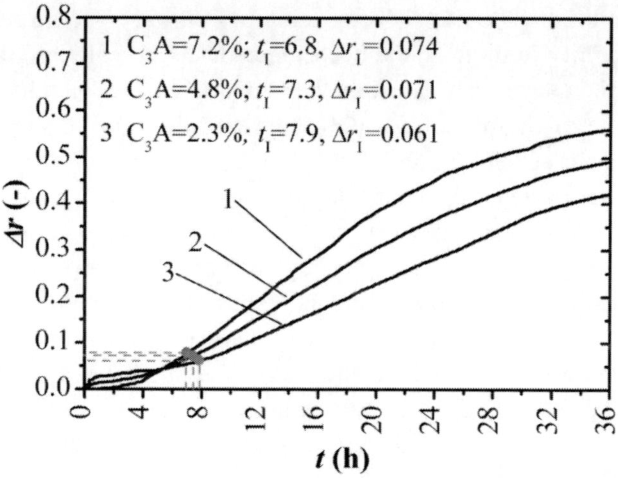

d)

Figure 7. Development of the Δr – t curves influenced by; a) w/c ratio, b) ambient temperature, c) cement fineness, d) amount of C_3A. Solid dots stand for the initial setting, t_I and Δr_I represent the initial setting time and corresponding Δr value, respectively.

Determination of Different Milestones during the Formation of Structure of Hardening Cementitious Materials Using USWR Method

Due to the unambiguous importance of initial and final setting times, a research has been also performed in order to estimate these essential characteristics with the use of the USWR method. According to the original research performed by Valič et al. (1999) both setting times were found to appear at the same level of Δr, independently of w/c ratio changing from 0.25-0.30. Sun et al. (2004) performed a similar study on three cement pastes with w/c ratios equal to 0.35, 0.50, and 0.60 and suggested to define a specific value of *RL* within a certain interval to indicate the occurrence of the initial setting regardless of w/c ratio. On the contrary, final setting cannot be defined by using a specific value of *RL*. According to Voigt et al. (2004) a consistent relationship between initial setting and the point at which the *RL* – *t* curve starts to increase could be identified. Using a L-wave pulse reflection technique, Ozturk et al. (2006) observed that initial and final setting times coincide with the time of the first inflection point on the curve representing the development of the shear wave reflection coefficient *WRF* in time and the time when the *WRF* – *t* curve starts to increase after having reached its minimum, respectively.

A comprehensive experimental work has been performed with the objective to establish a relationship between reflection loss *RL* and various microstructural parameters. Using Hymostruc model, a strong linear trend was found between the reflection loss and the degree of cement hydration (Sun et al., 2004, Sun, 2005) for cement pastes with different w/c ratios (0.35 – 0.60) cured under constant temperature of 25°C. Similar conclusions are presented by Voigt et al. (2005) in the case when different mortars were analyzed. However, this relationship is influenced by the w/c ratio where a low w/c ratio corresponds to a steep slope (Voigt et al., 2005). On the contrary, linear and hyperbolic relationships were established between the *RL* values and a decrease of capillary porosity and *RL* values and the contact area of solid phase, respectively. Interestingly, a bilinear relationship between the reflection loss and specific contact area was found in the case of mortars (Voigt et al., 2005).

According to Sun et al. (2004,) it is reasonable to define a specific reflection loss value within a certain probability in indicate the occurrence of the percolation of thc solid phase in the microstructure of the cement pastes where the w/c ratio can be ignored. Voigt et al. (2005) found out that the time of the percolation threshold and the time of increase of the *RL* – *t* curves are very similar for cement pastes with different w/c ratios.

Sun et al. (2006) investigated the possibility of using the USWT technique to monitor the viscoelastic behavior of Portland cement pastes with w/c between 0.40 and 0.60 cured under water at a constant temperature of 25°C. They indicated that even though both moduli develop after a very similar trend, the absolute value of a storage modulus calculated from the USWR measurements is greater than the one obtained from an oscillatory rheometric measurements via strain control. However, the shear modulus data calculated from results of the USWR method and resonant frequency method appeared to be almost equal. A strong linear trend was found between the viscosity values calculated from the USWR measurements and measured with a step rheometric method at various ages (Sun et al., 2006). Voigt et al. (2006) indicated that the time when the *RL* starts to increase from its initial constant value matches with the time when also the first significant increase of the compressive strength can be observed. They found out that compressive strength and reflection loss of mortar mixtures cured under different (isothermal) temperatures exhibit similar trends with age, both showing a well known cross-over effect (2004, 2006). An unvaried bilinear relationship was established between these two characteristics with the first part having a smaller slope than the second (Voigt, 2004). Later this relationship was simplified to a power law relationship (Voigt et al., 2006) independent of the w/c ratio, which was subdivided into three parts (Voigt et al., 2006, Voigt and Shah, 2004, Sun et al., 2005). A linear correlation between the shear wave reflection coefficient and compressive and bending strength was reported by Valič and Vuk (2000).

COMPARISON BETWEEN P-WAVE AND S-WAVES IN THEIR ABILITY TO MONITOR THE EARLY AGE HYDRATION PROCESS

According to Voigt et al. (2006), a direct comparison of the P-wave and S-wave velocity at early age allows important conclusions about the sensitivity of these two wave types to monitor the hydration process and formation of structure of cement based materials at early ages. Before the initial setting the P-waves were found to be more sensitive to the internal structure of different cementitious materials that the S-waves, originating from the fact that the S-waves are not influenced by the interparticle distance in the material and can only propagate if the solid phase is connected due to the cement hydration process (Voigt et al., 2006, Boumiz et al., 1996). This is in good agreement with the results presented by Trtnik et al. (2009a) who performed a comprehensive research with the objective to study the correlation between the ultrasonic wave transmission method (P-waves) and ultrasonic wave reflection method (S-waves) in their ability to monitor the setting process of various cement paste mixtures. They indicated that $\Delta r - v_P$ diagrams can be simplified into three characteristic phases with almost a horizontal line during the first phase. The end of the first phase could be closely related to the initial setting time of an arbitrary cement paste (Trtnik et al., 2009a), which further substantiate the assumptions that contrary to the P-waves, the differences in the internal structure of the cement paste do not influence the propagation of the S-waves. After the percolation threshold, S-waves were found to be more sensitive to the connectivity of solid particles than P-waves (d'Angelo et al., 1995). Voigt et al. (2006) indicated that S-wave velocity is governed by the same material property as the compressive strength, i.e. how well the individual particles are bonded together. According to Trtnik et al. (2009a), when the amount of solid phase reaches a certain value, the rate of increase of v_P values slows down while the values of Δr continue to increase appreciably. This indicates that S-wave are more sensitive to the microstructural changes during the period when the solid phase is fully connected and the pulse velocity follows the evolution of the total solid volume fraction.

CONCLUSION

The aim of this chapter was to demonstrate the ability of different ultrasonic methods in monitoring the formation of structure of various cementitious materials. Moreover, some important milestones during the early age setting and hardening process can be identified if properly considered. These US techniques are completely non-destructive, low-cost, and easy manageable and have therefore a great prospect of application in an every-day construction practice.

ACKNOWLEDGMENT

The work of G. Trtnik and G. Turk, presented within this state of the art, is financially supported by Slovenian Research Agency as a part of the research project No. Z2-4069. The support is gratefully acknowledged.

REFERENCES

Bentz, D. Modelling cement microstructure: pixels, particles and property prediction, *Mater. Struct.* 1999, 32, 187-195.

Boumiz A,; Vernet, C.; Cohen Tenoudji, F. Mechanical properties of cement pastes and mortars at early ages – Evolution with time and degree of hydration, *Adv. Cem. Based Mater., 1996*, 3, 94-106.

Chotard, T.; Gimet-Brert, N.; Smith, A.; Fargeto, D.; Bonnet, J. P.; Gault, C. Application of ultrasonic testing to describe the hydration of calcium aluminates cement at the early age, *Cem. Concr. Res.,* 2001, 31, 3, 405-412.

D'Angleo, R.; Plona T. J.; Shwartz, L. M.; Coveney, P. Ultrasonic measurements on hydrating cement slurries: Onset of shear wave propagation, *Adv. Cem. Based Mater.,* 1995, 2, 1, 8-14.

De Belie, N.; Grosse, C. U.; Kurz, J.; Reinhardt, H. W.; Ultrasound monitoring of the influence of different accelerating admixtures and cement types for shotcrete on setting and hardening behavior, *Cem. Concr. Res.,* 2005, 35, 11, 2087-2094.

Elvery, R. H.; Ibrahim, L. A. M. Ultrasonic assessment of concrete strength at carly ages, *Mag. Concr. Res.*, 1976, 12 , 181-190.

Gabrijel, I. Application of elastic waves for the analysis of setting and hardening in cement composites; *PhD Thesis;* University of Zagreb, Faculty of Civil Engineering; Zagreb, 2010.

Grosse, C. U.; Reinhardt, H. W.; Ultrasound Technique for Quality Control of Cementitious Materials, *Proceedings of the 15th Conference on Nondestructive Testing,* Rome, Italy, 2000.

Herb, A. *Indirekte Beobachtung des Erstarrens and Erhartens von Zementleim,* Universitat Stuttgart, Stuttgart, Mortel und und Betton mittels Schallwellenausbreitung, 2003.

Kamada, T.; Uchida, S.; Rokugo, K. Nondestructive evaluation of setting and hardening of cement paste based on ultrasonic propagation characteristics, *J. Adv. Concr. Technol.,* 2005, 3, 3, 343-353.

Keating, J.; Hannant, D. J.; Hibbert, A. P. Correlation between cube strength, ultrasonic pulse velocity and volume change for oil well cement slurries, *Cem. Concr. Res.,* 1989, 19, 5, 715-726.

Korhonen, C. J. *Off the Shelf Antifreeze Admixtures,* Cold Region Research and Engineering Laboratory, April, 2002.

Krauß, M.; Hariri, K. Determination of initial degree of hydration for improvement of early-age properties of concrete using ultrasonic wave propagation, *Cem. Concr. Compos.,* 2006, 28, 4, 299-306.

Lee, H. K.; Lee, K. M.; Kim, Y. H.; Yim, H.; Bae, D. B. Ultrasonic in-situ monitoring of setting process of high-performance concrete, *Cem. Concr. Res.,* 2004, 34, 4, 631-640.

Lokhorst, S. J. Deformational behavior of concrete influenced by hydration related changes of the microstructure, *PhD Thesis;* Delft University of technology, Delft; 1999.

Ozturk, O.; Krogel, P.; Grubl, J. S. Popovics. Improved ultrasonic wave reflection technique to monitor the setting of cement-based materials, *NDT&E Int.,* 2006, 39, 258-263.

Rapoport, J. R.; Popovics, J. S.; Kolluru, S. V. Shah, S. P. Using ultrasound to monitor stiffening process of concrete with admixtures, *ACI Mater J,* 2000, 97, 6, 675-683.

Reinhardt, H. W.; Grosse, C. U. Herb, T. Ultrasonic monitoring of setting and hardening of cement mortar – A new device, *Mater. Struct.,* 2000, 33, 233, 581-583.

Reinhardt, H. W.; Grosse, C. U. Continuous monitoring of setting and hardening of mortar and concrete; *Construction and Building Materials* 18, 2004, 145-154.

Robeyst, N.; Grosse, C. U.; De Belie, N. Relating ultrasonic measurements on fresh concrete with mineral additions to the microstructure development simulated by CemHyd3D, *Cem. Concr. Compos.* 33, 2011, 6, 680-693.

Robeyst, N.; Gruyaert, E.; Grosse, C. U.; De Belie, N. Monitoring the setting of concrete containing blast-furnace slag by measuring the ultrasonic p-wave velocity; *Cem. Concr. Res.,* 2008, 38, 10, 1169-1176.

Robeyst, N.; Grosse, C. U.; De Belie, N. Monitoring fresh concrete by ultrasonic transmission measurements: Exploratory multi-way analysis of the spectral information; *Chemometr. Intell. Lab.* 2009a, 95, 1, 64-73.

Robeyst, N.; Grosse, C. U.; De Belie, N. Measuring the change in ultrasonic p-wave energy transmitted in fresh mortar with additives to monitor the setting, *Cem. Concr. Res.,* 2009b, 39, 10, 868-875.

Sayers, C. M.; Dahlin, A. Propagation of ultrasound through hydrating cement paste at early times, *Adv. Cem. Based Mater.*, 1993, 1, 1, 12-21.

Schindler, A. K.; Dossey, T.; McCullough, B. F. Temperature control during construction to improve the long term performance of Portland cement concrete pavements; Texas Department of Transportation; *Research project No. 0-1700-2;* The University of Texas at Austin, 2002.

Stepišnik, J.; Lukač, M.; Kocuvan, I. Measurement of cement hydration by ultrasonics, *American Ceramic Society Bulletin,* 1981, 60, 4, 481-483.

Sun, Z. Monitoring the early-age properties of cementitious materials with ultrasonic wave reflection method at macro-and micro-structural levels; *PhD. Thesis;* Northwestern University, 2005.

Subramaniam, K. V.; Lee, J.; Christensen, B. J. Monitoring the setting behavior of cementitious materials using one-sided ultrasonic measurements, *Cem. Concr. Res.,* 2005, 35, 5, 850-857.

Subramaniam, K. V.; Wang, X. An investigation of microstructure evolution in cement paste through setting using ultrasonic and rheological measurements, *Cem. Concr. Res.,* 2010, 40, 1, 33-44.

Subramaniam, K. V.; Wang, X. Ultrasonic monitoring of capillary porosity and elastic properties in hydrating cement paste, *Cem. Concr. Compos.,* 2011, 33, 3, 389-401.

Sun, Z. Ye, G. Voigt, T. Shah, S. P. Van Breugel, K. Early age properties of Portland cement pastes investigated with ultrasonic shear waves and numerical simulations, J. Weiss, K. Kovler, J. Marchand, S. Mindess (ed.), *International RILEM Symposium on Concrete Science and Engineering,* RILEM Publications SARL, 2004.

Sun, Z.; Ye, G.; Voigt, T.; Shah, S. P., Van Breugel, K. Microstructural aspects of cement hydration – ultrasonic waves and numerical simulation, *5th international PhD symposium in Civil Engineering, Walraven, Blaauwendraad, Scarpas & Snijder* (ed.), Tylor and Franccis Group, London, 2004.

Sun, Z. Voigt, T. Shah, S. P. Temperature effects on the strength evaluation of cement-based materials with an ultrasonic wave reflection technique, *ACI Mater. J.,* 2005, 102, 4, 272-278.

Sun, Z.; Voigt, T.; Shah, S. P. Rheometric and ultrasonic investigations of viscoelastic properties of fresh Portland cement pastes, *Cem. Concr. Res.,* 2006, 36, 2, 278-287.

Trtnik, G.; Turk, G.; Kavčič, F.; Bokan Bosiljkov, V. Possibilities of using the ultrasonic wave transmission method to estimate initial setting time of cement paste, *Cem. Concr. Res.,* 2008, 38, 11, 1336-1342.

Trtnik, G.; Valič, M. I.; Kavčič, F.; Turk, G. Comparison between two ultrasonic methods in their ability to monitor the setting process of cement pastes, *Cem. Concr. Res.*, 2009a, 39, 10, 876-882.

Trtnik, G.; Kavčič, F.; Turk, G. Prediction of concrete strength using ultrasonic pulse velocity and artificial neural networks; *Ultrasonics,* 2009b, 49, 1, 53-60.

Trtnik, G. The use of ultrasonic method to monitor the setting and hardening process of concrete, *PhD. Thesis,* University of Ljubljana, Faculty of Civil and Geodetic Engineering, Ljubljana, 2009.

Valič, M. I. Hydration of cementitious materials by pulse echo USWR. Method, apparatus and application examples, *Cem. Concr. Res.,* 2000, 30, 10, 1633-1640.

Valič, M. I.; Vuk, T. A study of the influence of additives and admixtures on the hydration of cement with the pulsed USWR method, *Mater. Tehnol.*, 2000, 34, 5, 249.

Valič, M. I.; Stepišnik, J.; Gabrijelčič, M.; Vuk, T. Setting times measurements of Portland cements with pulsed USWR method, *KZT,* 1999, 33, 1-2, 83-86.

Valič, M.I.; Stepišnik, J. A study of cement pastes by reflection of ultrasonic shear waves. Part I: Apparatus, experimental method and application examples, *KZT*, 1998, 32, 6, 551-560.

Van Breugel,, K. Simulation of hydration and formation of structure of hardening cement-based materials, PhD Thesis, Delft, 1991.

Van der Widen, N. G. B. *Ultrasonic measurement for setting control of concrete,* International Workshop on Testing during Concrete Construction, Mainz, 1990.

Voigt, T. The Application of an Ultrasonic Shear Wave Reflection Method for Nondestructive Testing of Cement-Based Materials at Early Ages*, PhD. Thesis,* Leipzig 2004.

Voigt, T.; Malonn, T.; Shah, S. P. Green and early age compressive strength of extruded cement mortar monitored with compression tests and ultrasonic techniques, *Cem. Concr. Res.,* 2005, 36, 5, 885-866.

Voigt, T.; Grosse, C.; Sun, Z.; Shah, S. P.; Reinhardt, H. W. Comparison of ultrasonic wave transmission and reflection measurements with P- and S- waves on early age mortar and concrete, *Mater. Struct.,* 2005, 38, 282, 729-738.

Voigt, T.; Ye, G.; Sun, Z.; Shah, S. P.; Van Breugel, K. Early age microstructure of Portland cement mortar investigated by ultrasonic shear waves and numerical simulation, *Cem. Concr. Res.,* 2005, 35, 5, 858-866.

Voigt, T.; Sun, Z.; Shah, S. P. Health monitoring of early age concrete, *4th international conference on concrete under severe conditions: environment and loading,* Seoul, Korea, 2004.

Voigt, T.; Shah, S. P. Properties of early age Portland cement mortar monitored with shear wave reflection method, *ACI Mater J.,* 2004, 101, 6, 473-482.

Ye, G. Experimental study and numerical simulation of the development of microstructure and permeability of cementitious materials; *PhD Thesis;* Delft University of technology, Delft, 2003.

Ye, G.; Lura, P.; Van Breugel K.; Fraaij, A. L. A. Study on the development of the microstructure in cement-based materials by means of numerical simulation and ultrasonic pulse velocity measurement, *Cem. Concr. Compos.,* 2004, 26, 5, 491-497.

Ye, G.; Van Breugel, K.; Fraaij, A. Experimental study and numerical simulation on the formation of microstructure in cementitious material at early age, *Cem. Concr. Res.,* 2003, 33, 2, 233-239.

Ye, G.; Van Breugel, K.; Fraaij, A. L. A. Experimental study on ultrasonic pulse velocity evaluation of the microstructure of cementitious materials at early age, *HERON,* 2001, 46, 3, 161-167.

Zhang, Y.; Zhang, W.; She, W.; Ma, L.; Zhu, W. Ultrasound monitoring of setting and hardening process of ultra-high performance cementitious materials, *NDT&E Int.,* 2012, 47, 4, 177-184.

Zhu, J.; Kee, S. H.; Han, D.; Tsai Y. T. Effect of air voids on ultrasonic wave propagation in early age cement pastes, *Cem. Concr. Res.,* 2011, 41, 8, 872-881.

In: Ultrasonics: Theory, Techniques and Practical Applications ISBN: 978-1-62257-685-2
Editors: Hanako Ayabito and Mitsuko Katsukawa © 2013 Nova Science Publishers, Inc.

Chapter 3

USE OF UTRASONIC SPRAY-PYROLYSIS PROCESS: A BRIEF REVEW

L. Castañeda*

Escuela Superior de Ingeniería Mecánica y Eléctrica Unidad Ticoman del Instituto Politécnico Nacional, D. F., México

ABSTRACT

Ultrasonic spray-pyrolysis process has been utilized for the past forty years with great convenience for materials processing, especially for the fabrication of metal oxide thin films of semiconducting nature. The process renders several advantages associated to the adhesion, control of morphology, particle size, and homogeneity of thin films, apart from its low cost.

In this brief review, we present a short description of ultrasonic nebulization process of liquid through ultrasonic wave induced cavitation. Effects of different parameters like the ultrasound frequency, ultrasonic power, and fluid level in the atomization chamber on the nebulization rate of a liquid solvent of precursor have been discussed.

1. INTRODUCTION

Powder or film generation by aerosol processes involves a number of steps, or unit operations. For example, powder generation by spray pyrolysis involves atomizing a solution to form droplets, entraining the droplets in a gas stream to form an aerosol, transporting the aerosol to a heat source, heating the droplets to drive the reactions to form the products, collecting the powder, and cleaning up gaseous by-products. Similarly, powder generation by gasto-particle conversion involves delivering gaseous reactants to a heat source (the reactor), inducing the desired chemicals to react, collecting the powder that forms, and cleaning up gases. Film formation requires either the generation and deposition of droplets or the delivery of gaseous materials. In this chapter, we discuss the separate operations that together form an

*Phone: + (52) 555-729-6000, Fax: + (52) 555-729-6000; E-mail: luiscas@cablevision.net.mx.

aerosol powder or film generation system. Droplet generators, virtual and conventional impactors, cyclones, volatile precursor delivery systems, and powder collection by electrostatic precipitation and filtration are examined. For background, the reader is referred to Bayvel and Orzechowsk [1] and LeFebvre [2] for sprays; to Kodas and Hampden-Smith [3] for delivery al reactants; and to Willeke and Baron [4] for all other topics.

2. ATOMIZATION

2. 1. Introduction

The devices used to convert bulk liquids into droplets are usually called atomizers or nebulizers. Atomization is used in a variety of applications involving the generation and processing of fine particles. Liquid metal atomization is widely used to produce metal powders with diameters of 1 to 5000 μm. [5-7] Powders obtained by this route are used for the production of sintered machine parts, soldering materials, and metal pastes used to from patterns on circuit boards. Two starting points for information are the texts by Rayvel and Orzechowski [1] and Lefebvre. [2]

Atomization is also used to deposit anticorrosion coatings, plastics, resin, adhesives, and ceramic and stoneware glazing. It is utilized to wash and clean parts in Order to prepare them for subsequent physical or chemical processes in the microelectronics industry. Atomization is the basis of spray-drying in which liquids are sprayed into a heated gas stream to remove the moisture. The water or other liquid evaporates quickly because of the large surface area of the drops.

Atomization also plays a key role in spray pyrolysis (Figure 1.1) for powder production. In this case, droplet sizes are usually 0.1 to 100 μm, significantly smaller than those used for spray-drying. Various routes to films so exploit atomization. Droplet deposition relies on the generation of droplets of in volatile reactants. AACVD utilizes droplets of solutions of volatile precursors. (Figure 1.2).

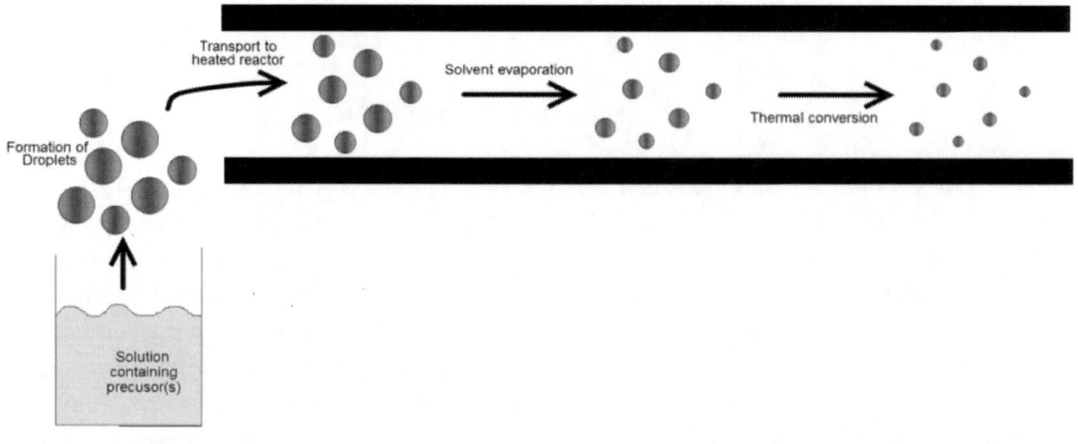

Figure 1.1. Overview of key step in spray pyrolysis.

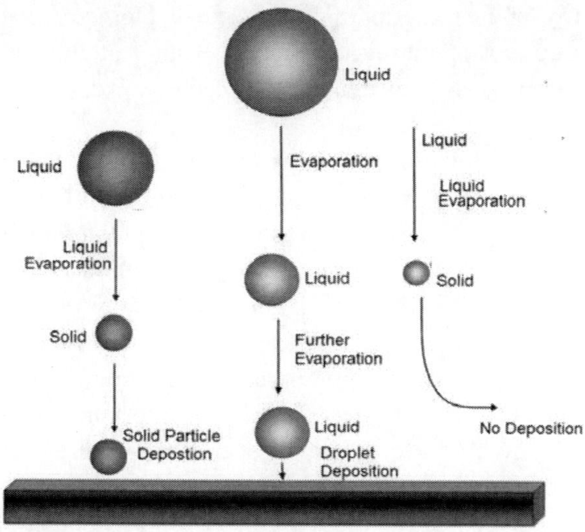

Figure 1.2. Droplet of different size can behave differently during deposition. Large droplets may deposit as droplets, while smaller may lose all solvent and form solid particles that may or may not deposit.

The requirements for the droplet size distribution and, therefore, the atomizer, vary, depending on the application and scale, laboratory or industrial. The discussion that follows outlines the differences among atomizers, describes their relevant characteristics, and explains which types of atomizers are suitable for the various aerosol routes to materials.

2. 2. Classification of Atomizers

Atomizers can be classified by means of various approaches, such as the type of energy used for atomization. The simplest class of atomizers uses the energy of the liquid itself to drive the breakup of the liquid; the pressure of the fluid is converted into kinetic energy, which atomizes the Iiquid. These atomizers can be classified into three types: *jet atomizers, swirl atomizers, and jet-swirl atomizers.* [1,2] Jet atomizers themselves can be grouped into two types: continuous and intermittent. Continuous atomizers are the simplest of all atomizers, while intermittent atomizers are the most complex. Jet atomizers produce jets that break apart as they leave the atomizer. High injection pressures are required to produce the kinetic energy necessary to break up the fluid into sufficiently small droplets. Swirl atomizers are the most widely used of all types of atomizers because they require lower pressures than jet atomizers. Swirl atomizers exploit the swirling of a liquid inside an atomizer to give a conical sheet of liquid that break up more easily than a jet. Jet-swirl atomizers combine the features of jet and swirl atomizers.

Pneumatic or gas-assist atomizers use the energy of a carrier gas to break up sheets or jets of a liquid. Pneumatic atomizers are the most complex of the atomizers because the liquid and gas phases interact in a complex manner. A common mode of interaction is the transverse introduction of a liquid stream into a high-velocity gas stream.

Rotary atomizers impinge a liquid onto a rapidly rotating surface, usually driven by electric motors that transmit its rotational energy to the liquid, causing it to leave the atomizer with high kinetic energy and break apart. This type of atomizer can provide narrow size distributions under the correct operating conditions. Rotary atomizers are often used in the atomization of solutions, because they with a liquid al low pressure. They are common in spray-drying systems.

Other types of atomizer, rely on more exotic phenomena to convert the liquid into droplets. Examples are *ultrasonic* and *electrospray atomizers.* These atomizers have been used extensively al the laboratory scale for aerosol routes to materials because of their ability to produce smaller droplets than can be obtained by other methods.

One of the major challenges for atomization technology is the production of high concentrations of droplets with small sizes. Most atomizer technologies e,g., jet, gas-assist, and rotary atomizers-when operated al industrial rates, produce droplets with diameters greater than 10 μm and concentrations lower than 10^7 droplets/cm^3. A variety of applications, including especially spray pyrolysis for powder generation, require droplets of 1 to 10 μm in size at concentrations greater than 10^7 droplets/cm^3 (but below the coagulation limit of roughly 10^9 droplets/cm^3). This has led to an emphasis on ultrasonic, electrostatic, and other technologies capable of producing droplets with sizes less than 10 μm at high concentrations.

The conventional jet, gas-assist, rotary, etc., atomizer technologic have been discussed in detail in various books. [1,2] For this reason, we focus our discussion un less conventional approaches, including ultrasonic and electrostatic atomizers.

2. 3. Pressure and Two-Fluid Atomization

In general, pressure and two-fluid atomizers cannot easily produce high concentrations of droplets smaller than 1 μm. Several designs are available for droplets in the 1 to 10 μm range, [8] and massive number of atomizers are available for droplets greater than 10 μm. We shall concentrate on the discussion of atomizers capable of producing droplets less than 10 μm in size.

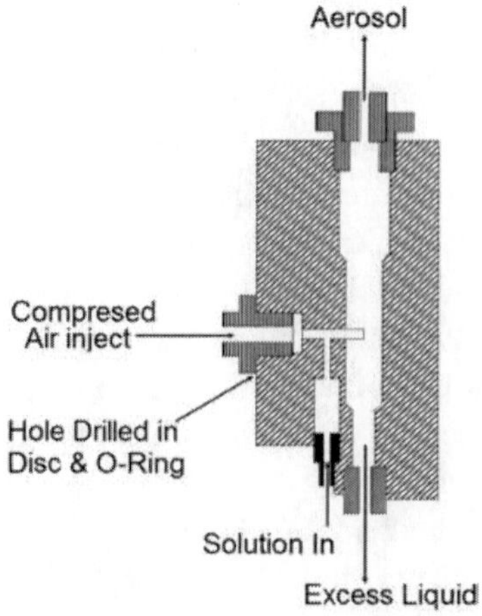

Figure 1.3. Collision-type aerosol generator capable of producing high concentrations of submicron-sized droplet.

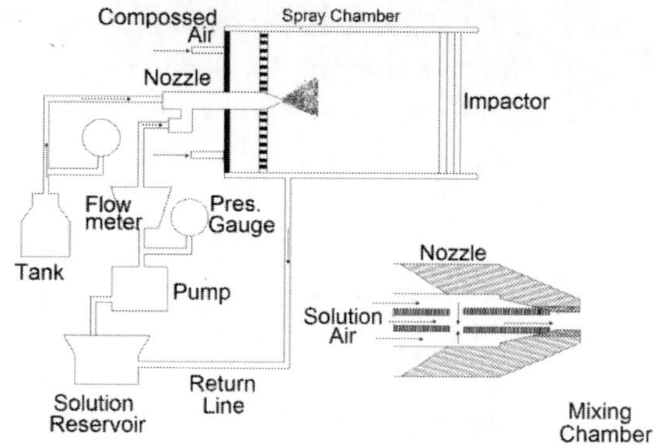

Figure 1.4. Effervescent spray system in which gas is dissolved in the liquid to be atomized and then forms bubble in the gas after exiting the nozzle, thereby breaking up the liquid stream into small droplets [11].

The physical phenomena that are exploited to produce droplets by these routes vary. *Pressure atomizers* work by discharging a liquid through a small aperture under high pressure. They have the advantage of not requiring any associated gas flow (therefore. high droplet concentrations can be achieved), but have the disadvantage of producing large droplet, sizes. *Rotary atomizers* work by centrifugal atomization: A liquid is directed onto a rotating element that converts the liquid stream to droplets. These atomizers, too, have the advantage of not requiring any associated gas flows, but have the disadvantage of providing large droplet sizes. *Air-assist atomizer* expose the liquid to a stream of air flowing al high Velocity. They require a gas flow to produce droplets, but have the advantage of providing somewhat smaller droplet sizes than are produced by pressure and rotary atomizer. *Air-blast atomizer* are similar to air-assist nozzles, but use much higher quantities of air or stream, these atomizers are discussed by Lefebvre and will not be examined here. [2]

Many variation of the *Collison* (not collision) design of produce droplets whit sizes on the order of 1 μm. [9,10] The Collison design relies on drawing a liquid stream in to a high-velocity gas stream that atomizes the liquid. The gas stream makes a sharp bend, and the larger droplets impact on the surface because of their greater inertia, while the smaller particles (with less inertia) follow the gas stream, thereby narrowing the size distribution and reducing the average droplet size. A typical design is shown in Figure 1.3. This particular atomizer produces droplets less than 1μm in size. Several designs (BGI. TSI) are available that produce droplet concentrations of 10^6 to 10^7 /cm^3 and sizes of 1 μm or less.

Effervescent atomization has been used for spray pyrolysis applications at the laboratory scale. This technique is illustrated in Figure 1.4. [11] The fluid to be atomized is pressurized to sufficient Ievels to dissolve significant quantities of the carrier gas in the liquid. The liquid is atomized into a region at atmospheric pressure in which the dissolved gas comes out of solution, resulting in further breakdown of the droplets. Droplet sizes are usually greater than 1 μm, but high throughputs are obtained relative to the Collision design.

2.4. Ultrasonic Atomizers

Ultrasonic atomizers use an ultrasonic transducer or horn that vibrates at ultrasonic frequencies (typically, 50 kHz-2.4 MHz) to produce the short wavelengths required for liquid atomization [12-19] the most useful characteristics of ultrasonic atomizers for materials processing by aerosol techniques are their low spray velocity the small or nonexistent amount of carrier gas required to operate them, and the small droplet size produced, which can be less than 10 μm. This makes it easy to carry small droplets at a high concentration into a flow reactor or towards a surface without using a high gas flow. Because of these useful characteristics, ultrasonic atomizers have been employed extensively in aerosol methods for generating both film and powders at the laboratory and on industrial scales. These types of atomizers are also used to spray acidic or basic solutions onto semiconductor wafers to etch features into them.

There are three classes of ultrasonic atomizers. One version passes the liquid through a tube that is driven ultrasonically (Figure 1.5). [20] In the other version, the liquid comes into contact with an ultrasonic horn, or the ultrasonic transducer is submerged in the liquid (Figure 1.6). [21] The principle of operation is similar for these examples of ultrasonic atomizers. When a liquid come into contact with an ultrasonically driven surface, a wave pattern appears on the surface of the liquid. When the amplitude of the vibration is sufficient, the wave height [14] is sufficient for the wave crests in the film to become unstable. This instability drives the formation of droplets, which are ejected from the surface. A carrier gas can then be used to sweep the droplets away from the surface.

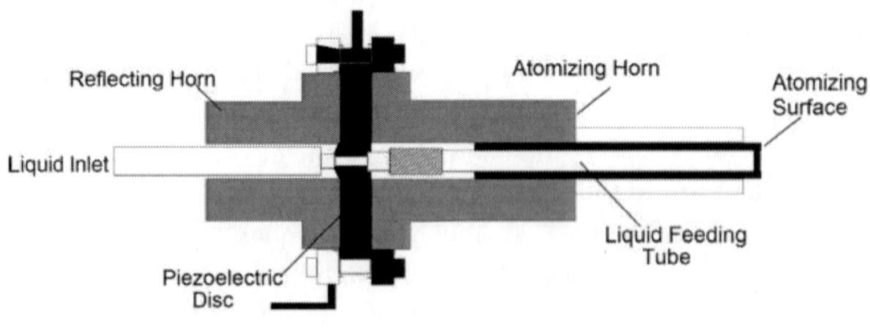

Figure 1.5. Ultrasonic horn atomizer in which a liquid stream flows through a tube and then comes into contact with a transducer (Courtesy of Berger) [20].

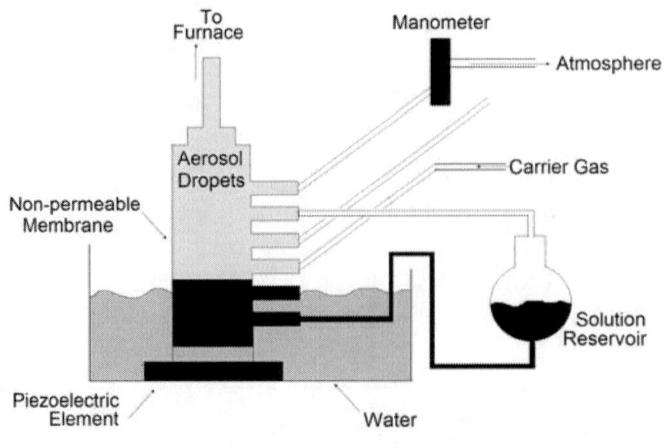

Figure 1.6. Submerged ultrasonic transducer for generation of droplets, The system shown call produce roughly 1 g/h of submicron-sized powder with a gas-phase particle concentration of 10^6 to $10^7/cm^3$. [21].

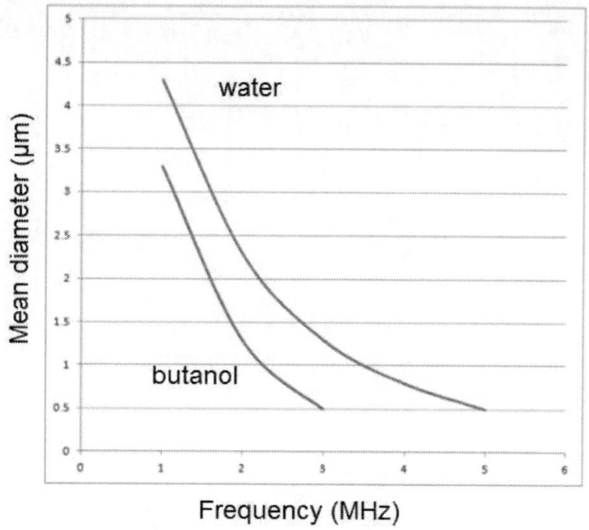

Figure 1.7. Calculated dependence of droplet size on transducer frequency for generation with submerge ultrasonic transducer.

The class of ultrasonic atomizers relying on a capillary tube has been described by Berger. [22] As shown in Figure 1.5, the nozzle consists of piezoelectric transducers placed between a pair of sections of the horn. The surface on which the droplets are formed is at the end of the stem, Liquid is delivered through a tube to the transducers. These atomizers which rely on making a liquid come into contact with an exposed transducer operate in a similar manner. The maximum flow rate is typically 10 liter/h at 55 kHz, whit a droplet or particle size range of 30 to 60 μm. The output drops dramatically as the frequency increases.

Ultrasonic transducers that are submerged in a liquid provide some of the smallest droplets of all ultrasonic atomizers. An example is shows in Figure 1.6. The submerged ultrasonic transducer can provide 1 to 5 μm droplets for frequencies of 1.5 to 2.4 MHz with low gas flow rates that correspond to high droplet concentrations. In medical applications, these droplets, when inhaled, can penetrate deep into the lungs to deliver medication. [19] The major disadvantage of this type atomizer is the low liquid flow rates per transducer.

Submerged ultrasonic transducers have been used in a large number of studies of spray pyrolysis. Much smaller particles can he obtained with these ultrasonic atomizer (typically, 2 to 4 μm) than with the other ultrasonic types, which produce droplets whit mass median diameters of 20 μm and larger. The droplet size depends on the surface tension, density, and frequency of the ultrasonic Source, according to the equation.

$$d_p = \left(\frac{\pi\sigma}{4\rho f_t^2}\right)^{1/3} \tag{1.1}$$

where d is the droplet diameter (cm), σ is the surface tension (erg/cm^2), ρ is the fluid density (g/cm^3), and f_t is the transducer frequency (1/sec).

Figure 1.7 shows how median droplet diameter depend of the frequency. [23] Transducers with frequencies ranging from 1.6 to 2.4 MHz are commercially available and produce droplets of 2 to 3 μm in median diameter.

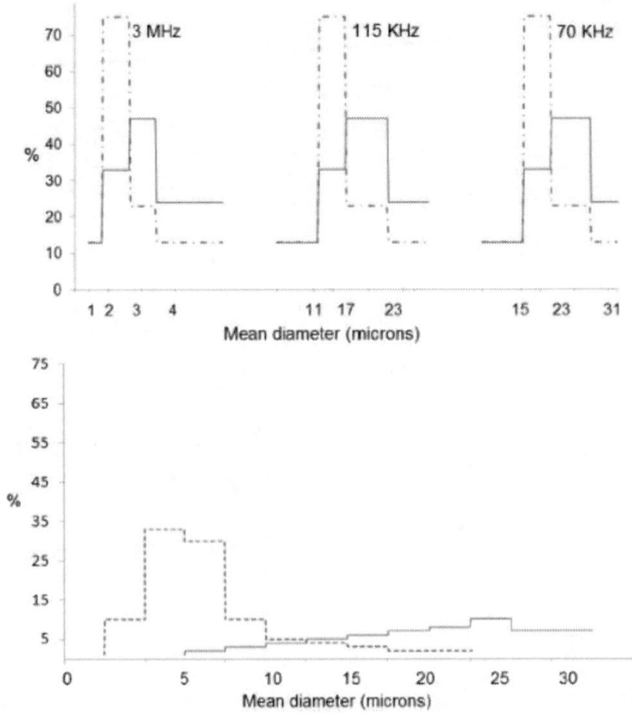

Figure 1.8. Number (- - -) and volume (----) distribution of water droplets obtainers by ultrasonic atomization (left) compared to number (- - -) and volume (---) distribution obtained by pneumatic atomization (right). Reprinted with permission from Langlet M., Joubert, J.C. In. Roa, C.N.R. (Ed.), Chemistry of Advanced Materials, 1993, Oxford, Blackwell Scientific, 55 [26].

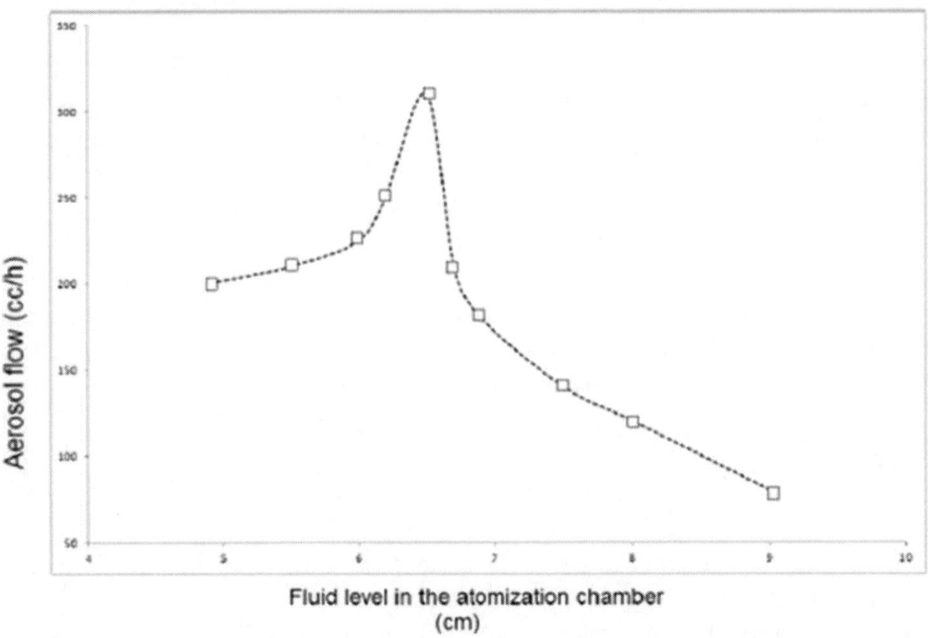

Figure 1.9. Measured dependence of aerosol flow as a function of level of liquid above transducer in system similar to that shown in Figure1.6. [26].

The primary advantages of ultrasonic generator whit submerged transducers are:

- Narrow droplet size distribution.
- Simplicity.
- High concentration of droplets.
- Droplet size appropriate for spray pyrolysis.
- Ability to vary droplet size by varying frequency.

a)

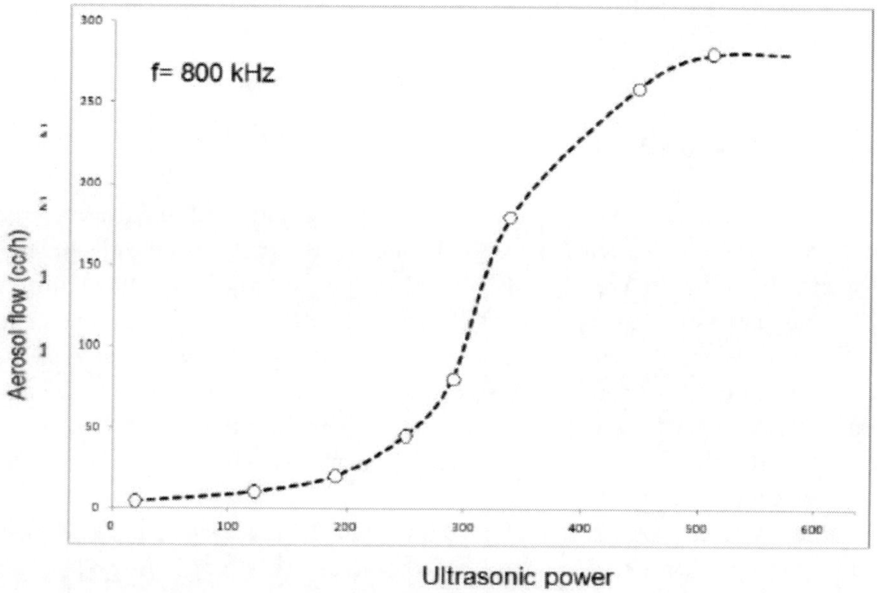

b)

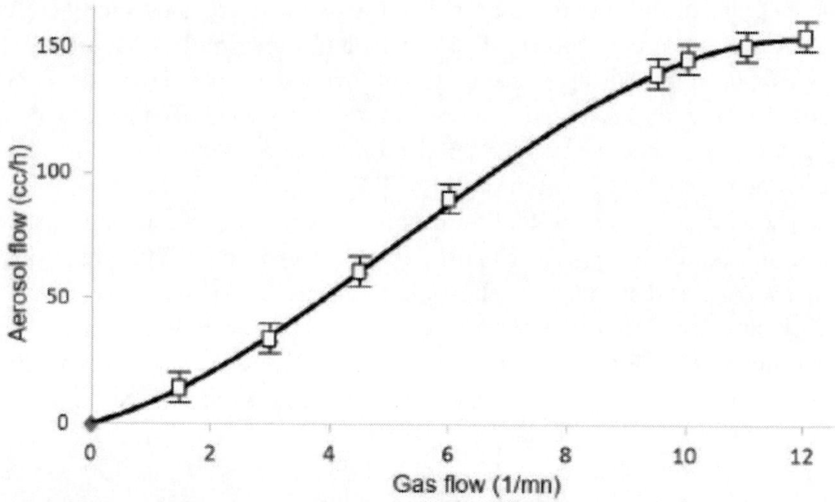

Figure 1.10. Effect of (a) ultrasonic power level and (b) carrier gas flow rate on ultrasonic atomization of butanol.

Several studies have measured droplet size distributions. [24,25] Figure 1.8 shows droplet size distributions produced using this type of ultrasonic generator. [26] The figure demonstrates that the droplet sizes required for most spray pyrolysis applications for powder generation correspond to frequencies near 2 MHz. Droplet concentrations depend on the details of operation of the reactor, but typical values are 10^6 to 10^7 droplets/cm^3.

Ultrasonic atomizers such as that shown in Figure 1.6 exhibit a dependence of the aerosol output on the level of the liquid above the transducer, as shown in Figure 1.9. [26] This requires careful design of these atomizers to maintain a constant Liquid level. Such generators also exhibit a strong dependence of the aerosol output on the ultrasonic power and gas flow. Typically, a certain gas flow is required to optimize the output in the sense that a maximum in aerosol concentration is obtained as the gas, flow rate is varied. This situation is illustrated in Figure 1.10. [26]

2.5. Electrostatic Atomizers

Electrostatic atomizers exploit charges for generating and modifying droplets and their sizes. There are two types of electrostatic atomization. In the first type, the electrostatic forces are insufficient to influence the stability of the droplet or liquid stream. However, once produced by conventional means, the drops are given electric charges by corona or induction charging. These atomizers are also termed hybrid atomizers because the liquid is atomized using pressure, pneumatic, or rotary atomizers and the drops are charged after being formed. A typical median diameter for the drops they produce is in the range 30 to 50 μm for a liquid flow rate of 50 mL/min. These atomizers find applications where high droplet deposition efficiencies are needed.

In the other type of electrostatic atomization, which has been studied extensively by aerosol scientists, the electrostatic forces by themselves are strong enough to atomize the liquid. Here the transfer of the charge to the liquid proceeds through contact (i.e., contact of the liquid with the source of electrons). This type of electrostatic atomization depends on the electrical conductivity of the liquid. The repulsion between charges on the liquid causes the stream to disintegrate into fine droplets. The expansion in the droplets due to the charges on the surface is opposed by surface tension force. Typical droplet sizes are submicron to millimeter, and the flow rates reported in the literature range from 10^{-7} to 1 mL/sec. Although the method is capable of producing submicron droplets, it does not yet allow simultaneous high liquid flow rates and droplets of that size [27-29]. As a result, it has been used primarily on the laboratory scale. A typical system for droplet generation is shown in Figure 1.11. [30] this type of atomizer has been used extensively to form droplets that are deposited onto surfaces. A measured size distribution is shown in Figure 1.12.

Electrostatic atomizers have the advantage that the charged droplet can be manipulated using electrical fields. This allows droplets to be deposited onto surfaces even in the presence of thermphoretics forces, which drive the particles away from a hot surface.

Submicron-sized metal powders have been produce by electrostatic atomization of molten metal. Particles as small as 0.01 μm and as large as 10 to 100 μm have been generated. A facility with a large number of such atomizers can produce 1000 kg of powder per day [32].

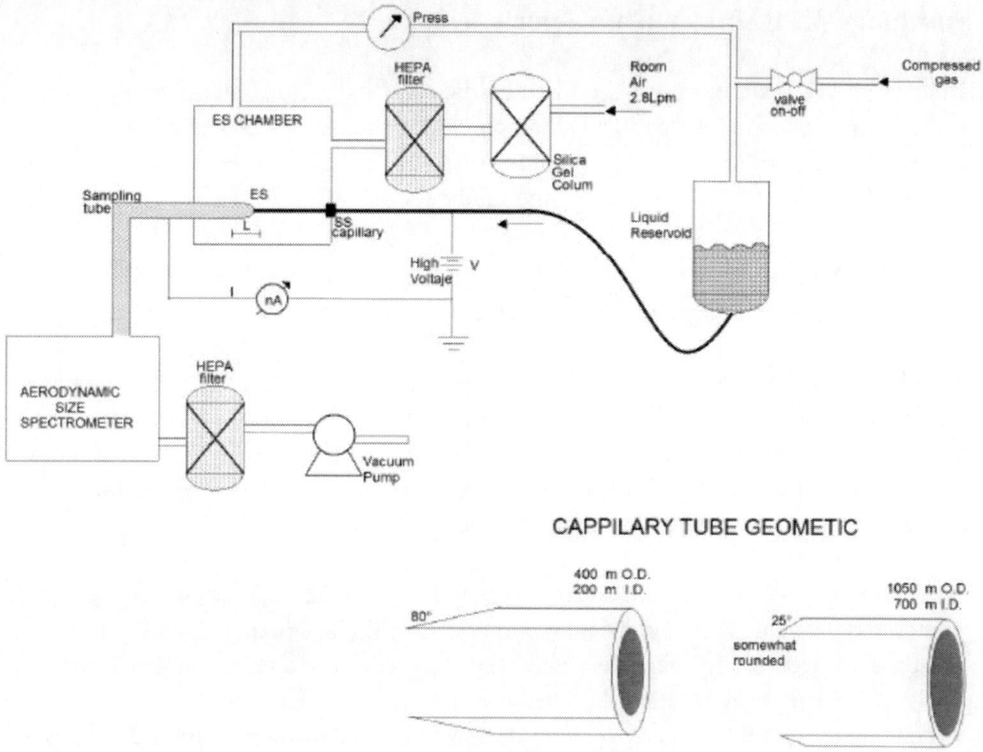

Figure 1.11. Electrospray system for droplet generation [30].

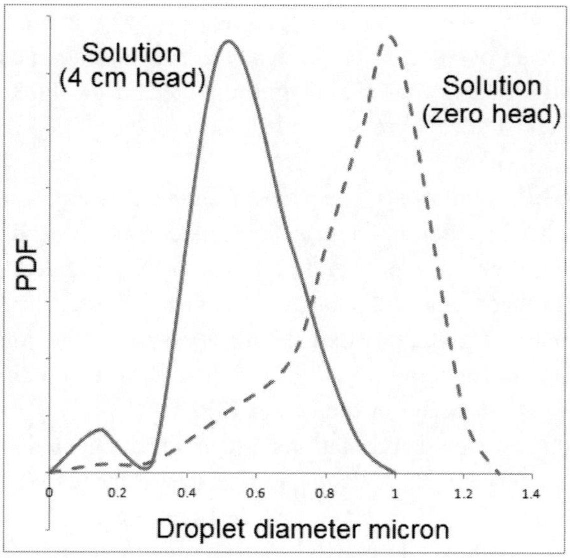

Figure 1.12. Droplet size distribution obtained from electrospraying [31].

2.1. Summary of Atomizer Characteristics

Table 1.1 gives a summary of the characteristic of the various atomizers. An important conclusion is that only the submerged ultrasonic, Collision-type, and electrospray atomizer are capable of producing droplets below roughly 10 μm. Other atomizers with similar droplet sizes, based on other exist but are not commercially available.

3. DRY POWDER FEEDERS

The dispersion of dry bulk powders into an aerosol stream has many applications, including inhalation studies, instrument calibration, aerosol characterization, drug delivery, and materials processing such as the classification of powder by particle size. The aerosolization of powders from bulk materials has been accomplished using fluidized beds and variation on power defers in combination whit jet mills and ventury-type powder dispersion devices. [33-35]

The optimum aerosol stream generated with these devices is once that contains unagglomerated particles with a consistent and controllable instantaneous and time-integrated particle mass concentration particles mass concentration and throughput These stream characteristics depend primarily on the attributes of the powder that is generated, the device used to generated, the device used to generate the powders, and the operating parameters of the generating device.

The dispersion characteristics of dry powders depend largely on the size, shape, and composition of the particles, the charging characteristics of the powder in the device, the extent of agglomeration of the particles in the powders, and the water absorption characteristics of the powder. In general, hydrophobic powders that are unagglomerated in the bulk disperse better into the powders, and the water adsorption characteristic of the powder. In general, hydrophobic powders that are unagglomerated in the bulk disperse into the gas than do agglomerated and hydrophilic powders, and dispensability increases with particle size.

Fluidized-bed particle generators use the mechanical energy of fluidized beads to disperse small, micron-sized particles. These generators consist of beads hundreds of microns in sizes that are loaded with the micron-sized particles to be generated. The mixture of beads and small particles is fluidized, and the small particles are released from the bed and carried off in the fluidizing gas. For a given powder, the mass output of the generator and the size of the particles generated are dependent upon the fluidizing gas the velocity, the bed height, and the mass particles relative to the mass of the beads. [33]

Powder-feeder-air-dispersion combinations use a mechanical dust-feeding mechanism such as a screw feeder or a scraper to feed the powder to a jet mill or venturi, where the particles are dispersed in a gas. For a given powder, the degree of dispersion and the mass output of the devices are dependent primarily upon the feed rate to the gas dispersion device and the gas flow characteristics in the device.

Table 1.1. Atomizer characteristics

Type	Average droplet Size	Spread of Distribution	Gas Flow Rate	Liquid Delivery Rate
Ultrasonic, submerged	1-10 μm	Narrow	Low	Low
Ultrasonic tube or horn	10-1000 μm	Medium	Low	Medium
Electrospray	<1-1000 μm	Narrow	Low	Low-high
Collision	<1-1000 μm	Broad	High	Low
Rotary	10-1000 μm	Broad	Low	High
Air blast	10-1000 μm	Broad	High	High
Air assist	10-1000 μm	Broad	High	High
Effervescent	10-1000 μm	Broad	Low	High

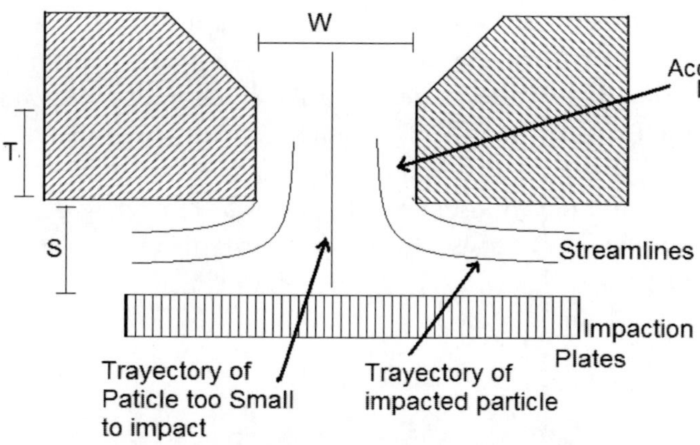

Figure 1.13. Impactor relying on inertial deposition of particles [36].

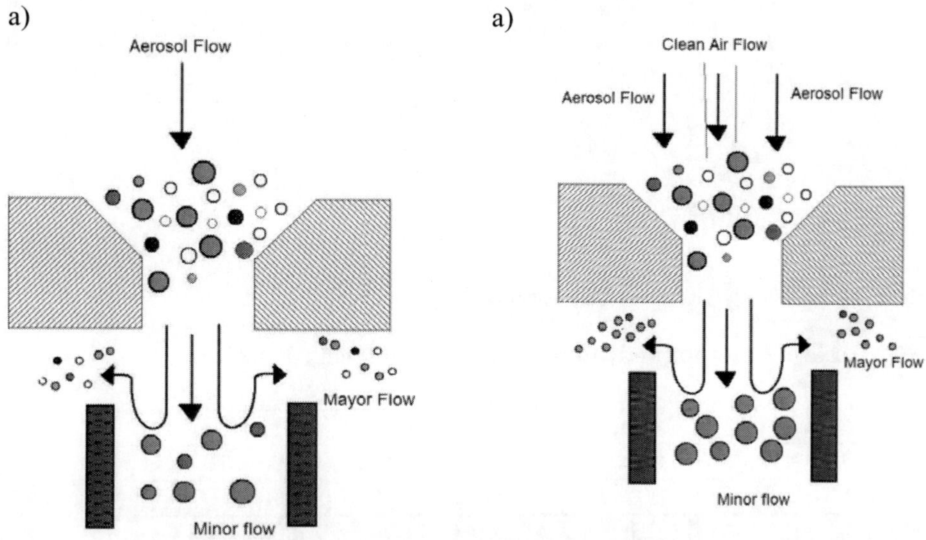

Figure 1.14. Virtual impactor relying on inertial separation of particles [47].

4. Droplet Impactors and Virtual Impactors

Inertial classification is a technique that can be used to remove larger droplets or particles from a gas stream by impaction on a surface or to remove smaller droplets or particle, while concentrating the remaining droplets or particles. For this reason, it is a powerful technique that has numerous applications in the processing of both films and powders. Inertial classifier, which include impactors, virtul impactors, and cyclones are widely used in the aerosol field, but also have seen limited applications in materials processing.

Droplet impactors (Figure 1.13) rely on directing a gas stream towards a surface near which the gas stream turns, and the inertia of the large particle or droplet causes them to collide with the surface. [36] This is also the arrangement used for fabrication of films by droplet deposition (spray pyrolysis of films). Droplet impactors have been used and studied both theoretically and experimentally. [37-44] The studies provide excellent and useful guidelines for the design of systems for film generation by droplet deposition. Many versions of these devices are available commercially for collecting particles, and many more have been designed, built, and used in special studies of materials processing by aerosol routes. Particles as small as 1 nm can be collected using this technique.

Virtual impactors (Figure 1.14) direct a gas stream towards an opening while drawing off most of the gas stream to the sides. [45-48] The larger particles are unable to follow the gas drawn off to the sides and therefore travel into the opening. As a result, the larger particles remain airborne and are concentrated in the small amount of remaining gas. The feature of keeping the larger particles in the gas phase is important for further processing of the droplets or particles. Cutoff points (the diameter above or below which particles are retained or removed) can be as small as 1 μm in practice.

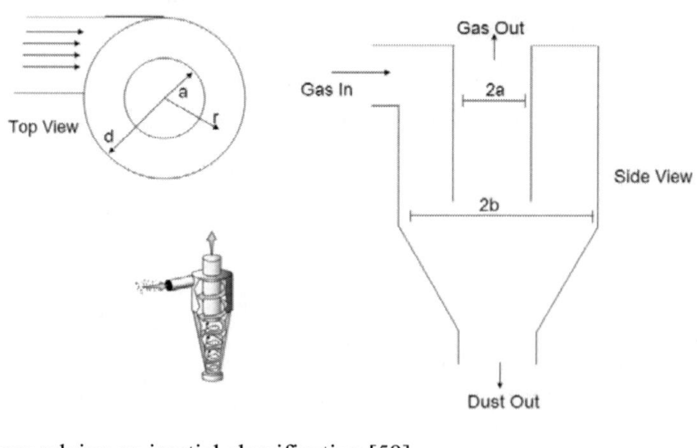

Figure 1.15. Cyclone relying on inertial classification [59].

Table 1.2. Stopping distance of particle for length scale of 10 cm

d_p , μm	Stopping Distance S (cm)	St
0.1	8.00×10^{-5}	8.00×10^{-6}
1	4.00×10^{-3}	4.00×10^{-4}
10	0.23	0.023
100	13.0	1.300

Cyclones (Figure 1.15) are also widely used to remove larger particles from a gas stream. [4,59] Cyclones rely on passing a gas tangentially into a cylindrical-shaped container, resulting in a gas flow in the radial direction. The inertia of the particles causes them to collide with the walls, where they are captured. Cyclones can collect much larger quantities of particles than impactors can and, for this reason, are often used to collect particles from gas stream. Numerous models are available commercially. Cyclones are most effective for particles larger than a few micrometers in size.

The inertial collection or particles relies on the momentum of the particles, Collection occurs by forcing the gas flow to make a sharp turn and then capturing the particles with sufficient inertia to cross gas streamlines that collide with the surface. Particles with less inertia remain in the gas flow and do not cross streamlines, A good example of inertial collection is an automobile moving down the highway. Insect do not usually hit the windshield at low speeds (relative to each other), but do so at high speeds. Large insects hit the windshield, while smaller insects do not. Finally, the bugs tend to strike region of the car where the flow of the air has to turn more sharply. This situation is quantified in terms of the Stokes number. Table 1.2.

$$St = \frac{\rho_p D_P^2 U C_c}{18\mu L} = \frac{S}{L} \quad (1.2)$$

where S is the stopping distance (the distance a particle travels in a gas until in reaches zero velocity whit respect to the gas), ρ_p is the particle density, d_p is the particle diameter, U is the velocity of the fluid, μ is the viscosity of the gas and L is a characteristics length of the flow geometry.

It is useful to compare the operating principles of the different types of inertial devices. A conventional impactor. (Figure 1.13) consists of a gas jet directed at a flat plate. Variations include either round or rectangular nozzles, single or multiple nozzles, and flat or cylindrical impaction surfaces. In a virtual impactor (Figure 1.14), the impaction plate is replaced by a collection tube, which is chosen to be larger than the jet. The larger particles travel nearly in a straight line into the collection tube, while the smaller particles leave with the majority of the gas flow. Only a small fraction (often just 10%) of the original amount of gas passes through the collection tube, thereby concentrating the particles. In a Cyclone, the aerosol stream is through an inlet and enters a cylindrical region tangentially at the outer edge of the cylinder, which is connected to a conical-shaped region. The gas spirals down the outer edge of the cylinder and cone walls and then reverses direction, after which it spirals upward along the center of the cyclone and exits through the top. The larger particles are collected on the cylindrical and conical walls by means of their inertial forces. Agglomerates of impacted particles fall off the walls to the bottom, where they are collected. A major difference between virtual and conventional impactor is that particles smaller than the cutoff size of the virtual impactor remain in the both the major and minor flows. The small particles in the large-particle fraction can be eliminated by providing a central core of clean air in the nozzle of the impactor. Besides being an inertial classifier, a virtual impactor can be used to concentrate particles larger than the cutoff size. Because these particles are concentrated in the minor flow, the concentration factor is equal to the total flow rate the minor flow rate. Concentration factors of 20 can be obtained.

5. POWDER COLLECTION: FILTRATION AND ELECTROSTATIC PRECIPITATORS

Separation of particles from gases is used for a variety of applications that consist primarily of the cleaning of gas for use (compressed air). The removal of particles in a process as waste of product, or the cleaning of the air in an enclosure (e.g. the removal of cigarette smoke from a room through filtration and air circulation units). The discussion in this section will concentrate on the removal of particles as a powder product from an aerosol process stream. The reader is referred to Willeke and Baron for an in-depth discussion. [4]

Aerosol particles can be separated from a gas using sedimentation, centrifugal force, inertia, interception, electrostatic precipitation, diffusion, and thermophoresis (Figure 1.16) [60] Although a large number of devices are available that use one or more of these mechanisms to collect particles from an aerosol stream, the most effective mechanism of powder collection depends primarily on the particle size distribution, the particle density, and the gas mean free path of the aerosol stream.

In selecting a method for collecting powder, one must consider many aspects of the aerosol stream, because it is these features that will determine the most efficient method of particle collection. Among such features are the average particle size, particle size distribution, particle density, gas mean free path, and gas composition, chemical and physical properties of the powders, particle morphology, and end use of the powder. Other matters to be considered in selecting a powder collection device are the collection temperature, the mass loading of the aerosol stream, the stream flow rate, the desired collection efficiency, whether the particles or gas are corrosive, and even the value of the powder relative to the cost of the unit, all of which are related to the economics of the process.

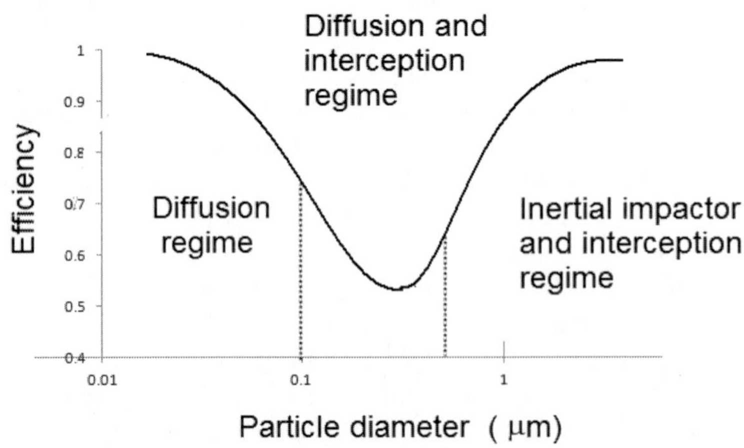

Figure 1.16. Filtration efficiency curve for particle in a gas. Note that the efficiency is high for small particles because of diffusion collection, in contrast to particle in liquid, where the collection efficiency is lower that smaller particle [60].

These factors determine the type, size, and materials of construction of the powder collection device, which, in turn, determine the cost of the unit. In general, the smaller the particles , the higher the collection efficiency, the more corrosive the environment, the higher the collection temperature, mass loading, and gas flow rate, the more the powder collection unit will cost. Of course, special consideration should be given to each application and device.

Devices used for particle collection include filters cyclones, settling chambers, electrostatic precipitators, thermophoretic collectors, scrubbers. impactors, and combinations of these devices. The most common type of collection devices are fillers, cyclones, settling chambers, and electrostatic collection devices, which are discussed in the remainder of this section.

5.1. Filtration

Filters can be designed for use in almost any application. For large-scale processes, filtration is generally more expensive than cyclones or settling chambers. Filters are commonly used, however, to collect particles with aerodynamic diameters less than 10 microns because these particles are difficult to collect using cyclones or settling chamber.

During filtration, particles are collected primarily by impaction, interception, and diffusion onto the filter substrate (Figure 1.16). A common misconception is that small particles are difficult to collect from gases. This is true only for liquids; Figure 1.16 shows that smaller aerosol particles are collected more efficiently than larger particles once the size of the particles is below a few tenths of a micrometer. This occurs because of the relatively high diffusion coefficients of particles in gases. Depending upon the type of filter and the aerosol used, particles can penetrate the top surface of the filter and enter into the filter body. For the collection of a product power this is undesirable, as powder that enter the body of the filter cannot be recovered easily. For this reason, high surface-loading filters are desirable when high recovery rates are required.

Filters can be used to process streams with flow rates ranging from less than 1mL/min to many thousands of cubic feet per minute. The size of the filter unit required depends on the system requirements, such as the flow rate and the maximum allowable pressure drop across the filter, the pressure drop per surface area of filter per gas flow rate, the mass loading of the aerosol, and the physical properties of the powder. The permeability of the filter depends upon the type of filter used and the pore size of the filter. For a given type of filter, the smaller the pore size, the lower is the permeability of the filter, and the larger is the surface area of the filter required for a given gas flow rate and pressure drop across the filter. Normally, filters with higher efficiency have lower permeabilities; therefore, the more stringent the collection efficiency requirements, the larger the filter unit will be.

Small laboratory processes normally use plate filter, small cylindrical filters, or small bag filters. Large industrial-scale processes use baghouse filters, which are large housings that contain multiple banks of bag or cylindrical filters. Many types of filters have different efficiencies, permeabilities, corrosion resistances, temperature resistances, and surface-loading properties, Some of the most commonly used filters for powder recovery on the laboratory scale are nylon, polysulfone, polycarbonate, polypropylene, polyethylene, metallic, ceramic and Teflon-based filters. A large number of baghouse filters also are available, ranging from simple polyester fabrics to stainless-steel mesh fillers and Teflon filters. In the removal of product powder, contaminates are undesirable; therefore, filters that do not flake off in to the powder are desired.

As the filter collects particles on the surface, the permeability of the filter decreases; thus, the pressure drop across the filter increases. As different powders load up in different configurations on the filter, the amount of material that can build up on the filter before the powder is removed depends on the powder's packing properties. In small-scale systems, once the filters are loaded with powder, they are normally removed and scraped to remove the powder, and then they are discarded. In large-scale processes, when the filters have been loaded with powder, the powder is normally discharged from the surface of the filter cartridges by shaking or purging the filter, from the inside out with gas that releases the powder, which then drops from the filters to a hopper, where it is collected in a bucket.

5.2. Cyclones

Cyclones operate on the principal of spiraling an aerosol gas through a cylinder, thereby imposing an artificial gravity on the particles, in a manner similar to a centrifuge, so that the particles are forced in an outward direction to be deposited on the inner wall of the cylinder. [4] Following deposition, the particles fall by gravity into the base of the cyclone, where they are collected in a pot. After the gas is spiraled through the centrifuge and the particles are collected, the gas exits the cyclone through the top of the cylinder.

Both droplets and dry particles can be collected using cyclones. These devices are most efficient, however, at collecting particles that have aerodynamic diameters greater than 5 microns, although specialty cyclones can collect particles as small as 1 or 2 microns in diameter. Al the same time, good collection efficiencies in this size range are difficult to obtain, especially in larger cyclone.

Cyclones are commercially available and can be designed to handle virtually any flow rate. They are perhaps the most economical and widely used powder collection devices available. Cyclones are normally constructed of metal and can be used at temperatures as high as 1000 °C. Advantages of cyclones over bag filters include less of a powder holdup, little or no increase in the pressure drop through the cyclone as powder is collected, simplicity in design and construction, the ability to process high-temperature gas streams, and relatively low cost. However, cyclone collection efficiencies for particles less than about 5 microns in diameter are often lower than those obtained using filters. Also, filters are usually much more forgiving in collection efficiencies for variation in flow rate.

Cyclones can be placed in series or in front of other collection devices to reduce the material loading of the gas. When cyclones are used in series, they act as size classifiers by removing smaller and smaller particles in each successive step.

5.3. Settling Chambers

Settling chambers use gravitational settling of particles as a means of collecting them. The aerosol stream is sent into the base of a relatively large chamber, which is sized such that the upward velocity of the gas is smaller than the settling velocity of the particles. The particles thus settle out of the gas. Due to their mode of collection, settling chambers are used for particles with significant settling velocities or particles generally greater than 25 to 50 microns. The size of settling chambers depends on the settling velocity of the particles and the flow rate of the gas being processed, settling chambers can be used for both large-and small scale processes.

5.4. Electrostatic Precipitators

In electrostatic precipitators, charged particles are collected on an oppositely charged surface (Figure 1.17). [34] this is accomplished by flowing an aerosol stream through an electrical field that ionizes the gas and creates a small electrical current from one plate lo another. The current exchanged between the oppositely charged surfaces depends primarily upon the composition of the gas, voltage potential between the charged surface and the distance between the surfaces. Normally the amperage between the plates is on the order of microamperes, and the voltage between the plates is tens of thousands of volts. If the gas breaks down, arcing between the charged surfaces will occur; this is not wanted because spikes in the amperage discharge are hard to deal with electrically, and the side receiving the discharge can develop pits. Also, the particle collection efficiency decreases drastically during arcing. For this reason, the operation characteristics of electrostatic precipitators are strongly dependent on the ionization potential of the gas.

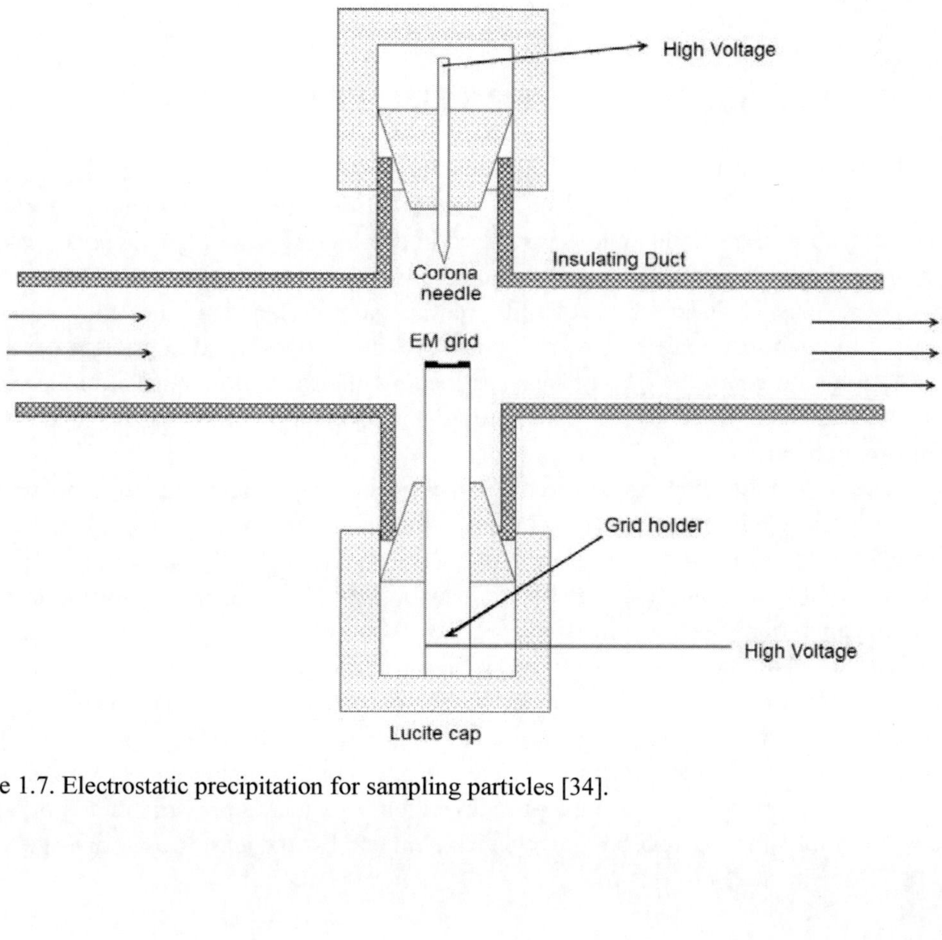

Figure 1.7. Electrostatic precipitation for sampling particles [34].

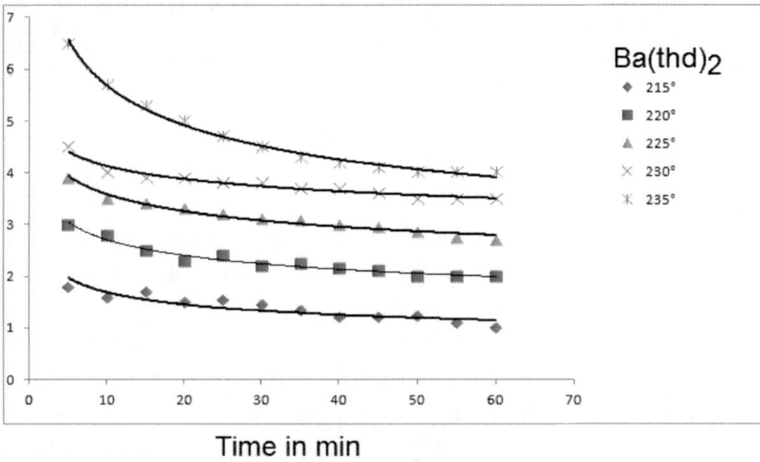

Figure 1.18. Evaporation rates of Ba(thd)2 as a function of time: P= 0 torr, Q = 34.65 sccm. Curves are drawn on te basis of fifted empirical equation [61].

One of the main advantages of the electrostatic precipitator is that the collection efficiency of the particles actually increases with a decrease in size, due to an increase in particle mobility with decrease size. [4] Electrostatic precipitators are commercially available at the scale of most processes.

6. Volatile Reactant Delivery Systems

The delivery of volatile precursor, to a reactor can be accomplished by a variety of methods, which are primarily determined by the volatility and the chemical and physics properties of the precursor. This topic is discussed in detail elsewhere. [3] In general, gaseous reactants are easier to deliver than liquid precursors which, in turn, are easier to deliver than solid reactants. This is due to the volatility of the gaseous reactants and the associated simplicity of the vaporization equipment. Generally, a continuous and consistent precursor output is desired and controlled precursor mass loading and mass flow rates are a necessity. Additionally, case and accuracy of determining and controlling the mass output of the precursor is beneficial.

Special considerations neck to be taken for each reactant. For example, reactants that are sensitive to certain environments (such as air or moisture) must be stored and delivered in environments that do not degrade the precursor. Also, the thermal stability of precursors can limit the time and temperature at which they can be heated. As an example, Figure 1.18 shows the evaporation rates of a solid precursor as a function of time at different temperatures. [61] The rate drops with time because of changes in precursor characteristics such as surface area and degree of degradation by solid-state reactions. In some instances, special techniques such as spraying are required to deliver thermally unstable precursors. [61]

Reactants that are already in a gaseous state such as saline or ammonia, can be delivered and regulated relatively easily by using a pressurized gas cylinder, a pressure regulator, and a flowmeter assembly that is resistant to chemical attack by the gas. Gases are the most convenient form of precursor to deliver, because they are easily delivered in a continuous and consistent stream that can be controlled and monitored.

Liquid reactants are normality generated using hobblers, evaporation pots, below-saturation-Level evaporation chambers, specialized liquid delivery systems, or spraying equipment such as nozzles or nebulizers. In standard bubblers, a carrier gas is bubbled through the liquid, thereby facilitating the evaporation of the precursor into the carrier gas. For a given precursor, the amount of precursor evaporated is dependent upon the temperature of the bubbler (which controls the vapor pressure of the precursor), the gas flow rate through the bubbler, the pressure of the bubbler (vacuum or pressurized), and the saturation level of the gas exiting the bubbler. Bubblers are perhaps the easiest and most common method of delivering volatile liquid precursors on a laboratory scale. The vapor production characteristics of bubblers are usually very good, and the mass output of bubblers can be easily monitored through changes in their volume or weight.

Evaporators are containers of the volatile liquid or solid precursors that are open to the system. Most often, evaporators are used on processes that operate under vacuum, and no carrier gases are used. However, evaporators can also be used in high-pressure systems if the precursor is volatile enough to deliver the reactant at the required rate. In evaporators, the precursors simply evaporate and expand into the system. The evaporation rate is a function of the precursor vapor pressure, the system pressure, and the evacuation rate of the system. These parameters can be controlled by varying the temperature of the evaporator and the system pressure. Quite often, the vapor pressure of the precursor sets the pressure of the system.

Liquid delivery systems are similar to evaporators, except that the precursor (usually a liquid) is delivered to the evaporation chamber in a controlled fashion. (The mass throughput of the liquid is controlled.) The evaporation chamber is operated at a temperature such that the precursor immediately evaporates. With this method, the precursor delivery rate is controlled by the feed rate to the evaporation chamber. A liquid delivery system can be used for pure liquids, solutions, or solids, although the delivery of liquids is much easier than that of solid reactants. Liquid reactants can be delivered to the evaporation chamber by a pump or a spray. A spray is often better because it facilitates more efficient evaporation. Solids can be delivered with a dry powder dispenser or by dissolving the precursor in an inner solution spraying the solutions into a chamber where it evaporates. Spraying solid precursor solutions is not always possible, because few solvents are compatible with both the precursor and the process itself. Still, the approach is valuable for precursor that are thermally unstable: The precursor does not have to be heated for extended periods of time and is heated and evaporated only as needed.

Solid reactants are usually the most difficult precursors to deliver at a consistent vapor feed rate. The simplest method for delivering solids is to flow gas over a heated bed of the solid reactant. However, the vaporization rate of the precursor can vary greatly, as the saturation level of the gas is dependent upon the surface area of the powder, which changes as it evaporates (Figure 1.18). One method of getting around this problem is to provide a large amount of powder surface area, so that the carrier gas always gets saturated in the evaporator or the surface area of the precursor powder changes very little with time. A method for accomplishing this is to provide a packed bed of the precursor and a high substrate surface area, such as beads, and then flow the carrier gas through the packed bed as in Figure 1.19) [62]. For this case, channeling through the bed must avoided.

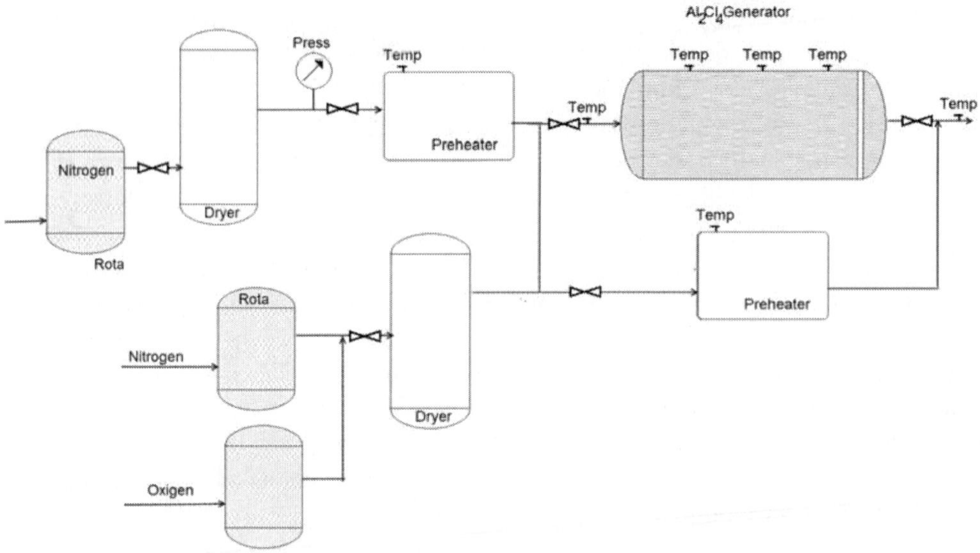

Figure 1.19. System for generating AlCI3, vapor for introduction into a gas-to-particle conversion reactor [62].

As discussed previously, solid reactants can also be delivered by spraying solutions or by using evaporators. Dissolving the solid reactant in a liquid and spraying the solution into an evaporation chamber is advantageous because the delivery rate is controlled and steady. However, as mentioned before, solvents that are compatible with the process and the precursor are not always available.

Finally, reaction-based vapor generators that take advantage of the reaction of two or more bulk materials to generate the desired volatile reactant are sometimes used. Usually, a solid and a gas are reacted to give off a volatile precursor, as is done in the production of $Ti0_2$, wherein raw rutileore or titanium metal is reacted with chlorine gas to form a gaseous $TiCl_4$, which is purified and delivered to be processed. In using this type of vapor generator, is often advantage to have the gas feed stream rate be limiting such that the delivery of the vapor precursor is regulated by the amount of gas fed into the generator.

7. Ultrasonic Spray Pyrolysis

In the ultrasonic spray pyrolysis system [63], a geyser is formed in the liquid surface by the action of a beam of ultrasonic vibrations which is directed to a liquid-gas interface. This geyser is followed by the generation of a spray formed by microscopic droplets. A carrier gas transports and directs the spray toward the surface of a hot substrate where the film formation is produced (see Figure 2.6).

7.1. Ultrasonic Atomization

The ultrasonic irradiation of a liquid with excitation potentials above a threshold value, results in the atomization of the liquid into microscopic droplets which forms a spray. The average diameter of the droplets generated by the ultrasonic beam (D) depends directly on the wavelength (λ_c) of the ultrasonic signal in the solution [63]:

$$D = \alpha \lambda_c \tag{1.3}$$

where α is a constant.

Using the Kelvin equation [64], the ultrasonic signal length in the solution can be written:

$$\lambda_c = \left(\frac{8\pi\sigma}{\rho f^2} \right)^{1/3} \tag{1.4}$$

where: σ represents the surface tension of the liquid, ρ the liquid density, f ultrasonic excitation frequency.

An experimental determination of the value of the constant α [65] is 0.34 which allows expressing the average droplet diameter as:

$$D = 0.34 \left(\frac{8\pi\sigma}{\rho f^2} \right)^{1/3} \approx \left(\frac{\pi\sigma}{4\rho f^2} \right)^{1/3} \tag{1.5}$$

The above equation shows clearly that the average diameter of the generated droplets (D) depends on the ultrasonic excitation frequency, and on the characteristics of the irradiated liquid (surface tension and solution density).

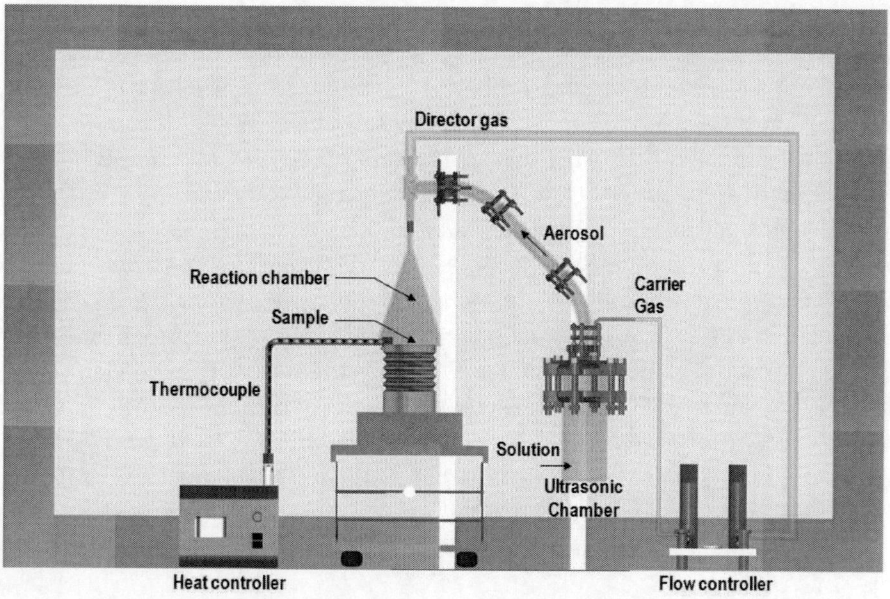

Figure 1.20. Schematic diagram of the ultrasonic spray pyrolysis method.

The main parameters in the ultrasonic atomization are [63]:

I. *Ultrasonic power*. Once generated over a threshold value, the amount of spray is continuously increasing as a function of the power applied to the ceramic. This parameter can control the amount of spray generated. Figure 1.10a shows the power effect of the ultrasonic source in the ultrasonic atomization of a butanol solution.
II. *Frequency of excitation*. The excitation frequency of the ceramic must be close to the resonance frequency of the molecules forming the precursor solution, this in order to optimize the atomization.
III. *Carrier and director gases*. The director gas leads the spray to the substrate in which the film is deposited. The spray flow rate is a function of the flow rate of the carrier gas. These parameters also allow control the rate of flow of spray[]. Figure 1.9 shows the effect of carrier gas flow in the ultrasonic atomization of a butanol solution.
IV. *Fluid level in the atomization chamber*. For each liquid being atomized and taking into account the geometry of the atomizing chamber prepared according to the geometry of the nozzle and its separation distance from the substrate (see Figure 1.20) a liquid level h_0 is calculated, which corresponds to the maximum amount of spray generated. Figure 1.9 shows the effect of the liquid level h_0 in the atomization chamber in the ultrasonic atomization of a butanol solution.
V. *Precursor solution*. As mentioned above the physical properties of the solution (surface tension and density of the solution) play an important role in the ultrasonic atomization model.

REFERENCES

[1] Bayvel, L. and Orzechowski. Z. *Liquid Atomization*. Washington, D. C: Taylor and Francis, 1993.

[2] Lefevre, A. H. *Atomization and Sprays*, New York: Hemisphere Publishing Co., 1989.

[3] Kodas, T. T. and 1994. Hampden-Smith, M. J. (Eds.). *The Chemistry of Metal C. V. D.*, Weinhein, West Germany: V. C. H., Verlagsgesellschaft, 1994.

[4] Willeke. K. and Bmoll. P. A. (Eds.). *Handbook of Aerosol Measurement: Principles, techniques and Applications.* New York: Van Nostrand Reinhold 1993.

[5] See, J. B. and Johnston. *G. H. Powder Technol*. 21, 1978, 119-133.

[6] Roylance, T. F., Pickering, S. J.., Hay, N., and Thomas, G. H, *A rotary cup air blast atomizer for molten slag- an investigation of particle size and spray trajectory. Institute of Energy*, ICLASS-85, 3rd International Conference on Liquid Atomization and Spray Systems at Imperial College, London, 8/9/10 July 1985. LP/VB 2. London: 1985, 1-16.

[7] Smith, N. P., Norris. J. O. W., Hemsley D. L., and Yeoman, M. L., *Measurement of the size and velocity of liquid metal drops during atomization. 3rd Int. Conf. on Liquid Atomzation and Spray Systerms*, Imperial College, London, 1985, LP/IVB(b)/ 1, London: 1985, 1 -9.

[8] Salisbury, C. L., Tuncel, G., and Ondov, *J. M. Aerosol Sci. Technol*. 15, 1991, 156-169.

[9] May, K. R. *J. Aerosol Sci*. 4, 1973, 235-243.

[10] Mercer. T. T. *Aerosol Technology in Hazard Evaluation*. New York: Academic Press, 1973, Chapter 9.

[11] Messing,G. *Nanoparticulates Workshop*, Monterey, C. A., 1994.

[12] Denton, M. B. and Schwartz. D. B. *Rev. Sci. Intrum*. 45(1), 1974, 81-83.

[13] Goddard, R. F., Mercer, T. T., O'Neill, P. X. F., Flores, R. L. and Sanchez, R. *J. Asthama Research* 5(4), 1968, 355- 368.

[14] Lang, R. J. *J. Acoustical Soc. Am.* 34(1), 1962, 6-8.

[15] Lierke, E. G. *Ultrasonic*, October 1967,214-218.

[16] Perron, R, R., *I. E. E. E. Trans. Sonics and Ultrasonics* SU-14(4), 1967, 149-153.

[17] Peskin, R. L and Raco. R. J. *J. Accoustical Soc. Am*. 35(9), 1963, 1378-1381.

[18] Pointon, A. J. *I. E. E. E. Proc.* 129 pt. A, 1982. 285-307.

[19] Topp, M. N. and Eisenklam, *P. Ultrasonic* 1972, 95-101.

[20] Berger, H. L. *Characterization of a Class of Widely Applicable UltrasonicNozzle. Proc. 3rd Inter. Con. On Liquid Atomization and spray systems*, London, July 1985, 1A/2/1-13.

[21] Lyons S. W. *Aerosol dynamics of superconducting ceramic power Generation by Gas-Pase Process*. Master´s thesis, university of New Mexico, 1992.

[22] Berger, H. L. *Proceeding of the 3rd Int conf. on liquid atomization and Spray System A*. 2. London 1985, 1-13.

[23] Tang Q. *Aerosol-deposited YBaCuO thin films and superconducting optical detector*. Ph. D. dissertation, Twente University, Enschede, the Netherlands, 992.

[24] Rodes, C., Smith, T., Crouse, R. and Ramachandran, G. (1990) *Aerosol Sci. Technol.* 13, 220-229.

[25] Tarr, M. A., Zhu, G., and Browner, *Applied Spectroscopy,* 45(9), 199, 1424-1432.

[26] Langlet M., Joubert, J. C. In. Roa, C. N. R. (Ed.), Chemistry of Advanced Materials, 1993, Oxford, U. K. Blackwell Scientific Publication, 1993,55.

[27] Graf, P. E. Breakup of small liquid volume by electrical charging. *Proceedings of A. P. I. Research Conference on Distillate Fuel,* New York: Am. Petroleum Institute, 1710. 1962, CP62-4.

[28] Luther, F. E. Electrostatic automization of No. 2 heating oil. *Proceedings of A. P. I. Research Conference on Distillate Fuel, New York: Am Petroleum Institute*, 1710. 1962, CP62-3.

[29] Meesters, G. M. H., Vercoulen, P. H. W, Marijnissen, J. C. M., and Scarlett, B. *J. Aerosol Sci.* 23(1), 1992,37-49.

[30] Rosell-Lompart, J. and Fernandez de la Mora, J. *J. Aerosol Sci.* 25(6), 1994, 1093-1119.

[31] Naqwi, A. A. *J. Aerosol Sci.* 25(6),1994, 1201-1211.

[32] Bailey, A. G., *Atomization Spray Technol*. 2, 1986, 45.

[33] Carpenter, R. L. and Yerks, K. *Am. Ind. Hyg. Assoc. J.* 41, 1980. 449-454.

[34] Hínds, W. C. *Aerosol Technology,* New York: Wiley Interscience, 1982.

[35] Koch, W., Lodding, H., Oenning, G., and Muhle, H. *J. Aerosol Sci.* 17(3), 1986. 499-504.

[36] Marple, V., Rubow, K., and Olson, B. In Willeke, K. and Baron, P, A. (Eds.), *Handbook of Aerosol Measurement: Principles, Techniques and Application.* New York: Van Nostrand Reinhold, 1993.206.

[37] Biswas. P., Jones, C. L. and Flagan, R..C. *Aerosol Sci. Tech.* 7(2), 1987, 231-246.

[38] Burton, R. M. and Lundgren, D. A. *Aerosol Sci. Tech.* 6, 1987, 289-301.
[39] Her, J. Y. and Kim, S. S. *Aerosol Sci, Tech.* 14,1991, 157-164.
[40] Hillamo, R. E. and Kauppinen, E. l. *Aerosol Sci. Tech.*14(1). 1991, 33-47.
[41] Marple, V. A. and Olson, B. A., *Aerosol Sci. Tech.* 22, 1995, 124-134.
[42] Merecer, T. T. and Stafford, R. G., *Am. Occup. Hyg.* 12, 1969,41-48.
[43] Ranz, W.-E. and Marshal W. R., Jr. *Chemical Engineering Progress* 48 (4), 1952, 173-180.
[44] Regtuit, H. E., DeRuiter, C. J., Vrins, E. L. M., Hofschreuder, P., Oeseburg F., and Bemschop F. M. *J. Aerosol Sci.* 21 (7),1990,919-923.
[45] Anderson, T. L. Charlson, R. L. and Covert, D. S. *Aerosol Sci. Technol.* 19, 1993, 317-329.
[46] Biswas P. and: Flagan. R. C. *J. Aerosol Sci.* 19(1), 1988, 113- 121.
[47] Chien, H. and Lundgren, D. A. *Aerosol Sci. Technol.,*18.1993,376-388.
[48] Chen, B. T, Yeh, H. C. and. Cheng, Y. S, *J. Aerosol. Sci.* 16(4), 1985, 343-354.
[49] Chen, B. T., Yeh, H. C. and. Johnson, *J. Aerosol. Sci.* 16(4), 1985, 343-354.
[50] Forney, L. J., Revernhall, D. G., and Lee, S. S. *Environ. Sci. Technol.* 16(8), 982, 492ff.
[51] Friedman., W. and Willeke, K. *Aerosol Sci. Technol.* 5, 1986, 447-457.
[52] Keskinen, J., Janka, K., and Lehtimak, M. 6, 1987, 79-81.
[53] Novick, V. J. and Alvarez, J. L. *Aerosol Sci. Technol.* 6.,1987, 63-70.
[54] Pillai, R. S., Yeates, D. B., Eljamal, M., Miller, I. F.,and Hickey, A. J. *J. Aerosol Sci.* 25(1), 1994, 187-197.
[55] Rader, D. J. and Marple. V. A. *Aerosol Sci. Technol.* 4, 1985, 141-156.
[56] Ravenhall, D. G., Forney, J. L., and. Hubbard, A. L. *J. Colloid. Interface Sci.* 85(2), 1982,508-520.
[57] Sioutas, C., Koutrakis, P., and Burton, R. M. *J. Aerosol Sci.* 25(7), 1994,1321-1330.
[58] Sioutas, C., Koutrakis, P., and Olson, B. A. *Aerosol Sci. Technol.* 21, 1994,223-235.
[59] Friedlander, S. K. *Smoke, Dust and Haze*. New York: Wiley Interscience, 1977.
[60] Lee, K. W. and Ramamurthi, M., In Willeke, K. and Baron, P. A. (Eds.), *Handdoook of Aerosol Measurement: Principles, Techniques and Applications.* New York: Van Nostrand Reinhold, 1993, 179.
[61] Chou, K. S. and Tsai, G.·J. *Thermochim. Acta*, 240, 1994,129.
[62] Powell, Q. H. *Gas-phase production and coating of titiania in tubular, hot-wall reactors*. Master´s thesis, university of New Mexico, 1995.
[63] M. Langlet and J. C. Joubert, *The Pyrosol Process or the Pyrolysis of an Ultrasonically Generated Aerosol Chemistry of Advanced Materials,* 1973, Edited by C. N. R., R. A. O., F. R. S.
[64] J. W. S. Rayleigh, *The Theory of Sound, Volume Two*, Second Edition Revised and Enlarged, 1945, Dover Publications, New York, US.
[65] R. J. Lang. *J. Acoust. Soc. Am.*, 34(1962)6.

In: Ultrasonics: Theory, Techniques and Practical Applications ISBN: 978-1-62257-685-2
Editors: Hanako Ayabito and Mitsuko Katsukawa

Chapter 4

ULTRASONIC ASSISTED FIXED ABRASIVE MACHINING OF HARD-BRITTLE MATERIALS

Yongbo Wu,[1] Yaguo Li,[1] Jianguo Cao[1] and Zhiqiang Liang[2*]
[1]Department of Machine Intelligence and Systems Engineering, Akita Prefectural University, Yurihonjo, Japan
[2]Key laboratory of Fundamental Science for Advanced Machining, Beijing Institute of Technology, Beijing, China

ABSTRACT

Ultrasonics has been identified as a practical and vital technique in manufacturing, especially in the high efficiency and precision machining of difficult-to-machine materials, which, in some cases, is otherwise considerably difficult to be accomplished without ultrasonics. Traditional machining in the presence of ultrasonics can be categorized into ultrasonic machining (USM) and ultrasonic assisted machining (UAM). The USM involves the use of ultrasonics to energize loose abrasives and the energized abrasives impinge upon surfaces of workpieces to remove material in the form of mechanical actions such as hammering, rolling and sliding of abrasives and occasionally chemical actions- ultrasonic cavitation erosion. In contrast with USM, UAM refers to machining that ultrasonics is incorporated into widespread machining techniques such as cutting and grinding. The UAM has proven to enhance the machinability of difficult-to-machine materials and/or improve the quality of workpiece surfaces, which is due greatly to exceptional ability to dispel machining debris and/or to reduce machining forces and resultant machining heating.

This chapter concentrates on the introduction of the UAM technique where ultrasonics is applied to fixed abrasive machining as there is an increasing tendency to supersede loose abrasives owing mainly to low-efficiency and cost-ineffectiveness of loose abrasive machining. The chapter commences with a succinct introduction of the

* Yongbo Wu, Yaguo Li, Jianguo Cao: Department of Machine Intelligence and Systems Engineering, Akita Prefectural University, Yurihonjo 015-0055, Japan. Zhiqiang Liang: Key laboratory of Fundamental Science for Advanced Machining, Beijing Institute of Technology, Beijing 100081, China.

principles of the USM and the UAM and then proceeds to internal grinding with the aid of ultrasonics, which is followed by ultrasonic assisted grinding of mono-crystalline silicon and sapphire, and ends with a recently proposed technique of ultrasonic assisted fixed abrasive polishing of optical glass. In each part, processing principle and apparatus will be described concisely and the machining methods are ultimately evaluated in terms of surface roughness and/or material removal rate, etc.

1. Introduction

Ultrasonics has been recognized not only as a viable and dependable tool to prognosticate diseases in medical practices but also a prevailing technique in non-destructive detection and manufacturing. In manufacturing, ultrasonics has been extensively applied to difficult-to-machine materials. Traditional machining in the presence of ultrasonics can be delineated into ultrasonic machining (USM) and ultrasonic assisted machining (UAM).

1.1. Ultrasonic Machining (USM)

The application of ultrasonic vibration to loose abrasive machining of hard-brittle materials dates back to as long ago as 1920s [1], which was proposed for the purpose of fulfilling the ever-increasing needs for machining difficult-to-machine materials productively. This machining method is well-known ultrasonic machining (USM), in which a generator transfers electric energy to magnetostrictive or piezoelectric transducer which vibrates a machining tool (Figure 1). The tool attached to the transducer through a mechanical amplifier, vibrates at a frequency of over 20 kHz and a amplitude less than tens of μm along its own axis. The vibration, in turn, imparts a high velocity to abrasive grains by means of fluid media between the tool and the workpiece. In typical processes, the tool is separated from the workpiece surface and a gap/clearance between the tool and the workpiece surface is impregnated with fluids in which appropriate abrasives are dispersed. The material is removed by the energized abrasives, which invade the surface of workpiece [2]. The mechanism of material removal is believed to be due to mechanical abrasion by rolling, sliding, direct hammering of the abrasives, micro-chipping [3]. Chemical ultrasonic cavitation erosion also contributes to material removal [4]. The material removal rate (MRR) is elevated by, among others, static load, vibration amplitude, and vibration frequency. USM is ideally suited for drilling and holing hard-brittle materials. The USM generates no heating and induces no additional damage and is a stress-free machining method. However, the MRR in USM is quite low and the erosion of machining tool is a knotty problem.

1.2. Ultrasonic Assisted Machining (UAM)

The other application of ultrasonics to machining is to utilize ultrasonics as an auxiliary to supplement conventional prevailing machining techniques, e.g., cutting and grinding, which is referred to as ultrasonic assisted machining (UAM). Given its universality, ultrasonics can be incorporated into most machining techniques to implement hybrid machining. The ultrasonic assisted turning, cutting, and grinding are representative of the UAM. A typical UAM process, ultrasonic assisted surface grinding, is schematically illustrated in Figure 2. The material removal mechanism of UAM differs from that of USM. The material cannot be removed without ultrasonics, that is, the ultrasonics is indispensable in USM, while in UAM ultrasonics is a supplement to facilitate conventional machining techniques.

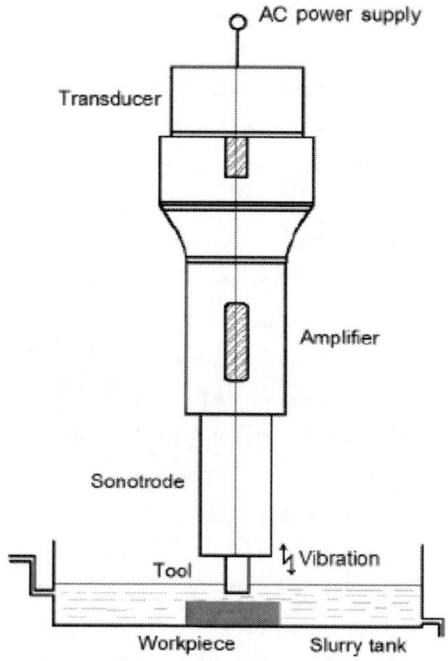

Figure 1. The alternate current (AC) is converted into vibration via a transducer, which is then amplified by an amplifier. The whole vibration generating gadget is referred to as a sonotrode. The machining tool is attached to the sonotrode. The slurry can be retained in a reservoir or circulated during machining.

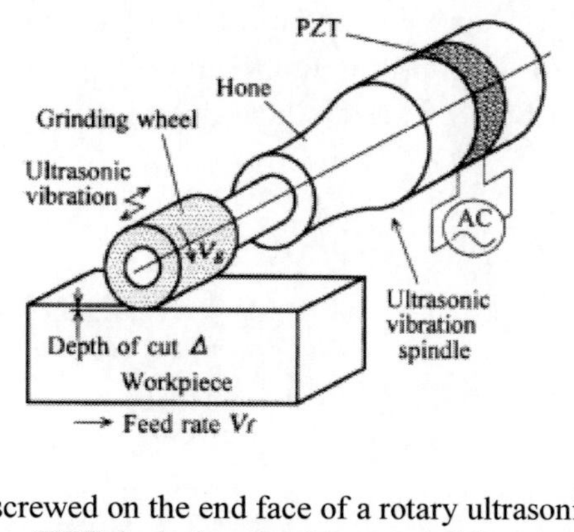

Figure 2. A grinding wheel is screwed on the end face of a rotary ultrasonic spindle composed of a hone and a bolted langevin transducer (PZT device) and rotates around its own axis. The electrical energy (AC voltage) is converted into vibration energy by the transducer and transmitted through the hone to the grinding wheel, resulting in a vibration of the grinding wheel at ultrasonic frequencies in its axial direction. Giving a depth of cut, Δ, between the workpiece and the grinding wheel, and feeding the workpiece right-forward at a feed rate of V_f results in an ultrasonic assisted surface grinding.

The process can be combined effectively with many conventional turning, drilling, and boring operations for both metallic and nonmetallic materials to fulfill such extraordinary manufacturing results as increased material removal rates, decreased machining forces, reduced tool wear, and improved surface finish. Recent researches confirm further that applying ultrasonic vibration can increase the critical depth of cut at which brittle-ductile transition arises [5, 6]. The increase in critical depth of cut is attributed to reduction in machining forces [5, 7].

Moreover, wear of tools, surface roughness of machined workpiece, surface integrity and MRR are improved as well [8-10]. Most of conventional UAM processes refer to the vibration of tools and/or abrasives while a method involving the vibration of workpiece in lieu of the vibration of tools/abrasives is presented by Egashira and Masuzawa to drill micro-holes [11]. More recently, the UAM has been extended into the generation of structured surface because applying ultrasonics can alleviate the burring and smooth the edge of microstructures as well as reduce the wearing of the machining tool.

Among the aforementioned UAM processes, those employing a tool with fixed abrasive either impregnated or electroplated on the tool surface can be performed at a constant pressure or a given depth of cut between the tool and the workpiece. This kind of UAM process is named ultrasonic assisted fixed abrasive machining (UAFAM). Our group practices the UAFAM to machine ceramics, silicon and sapphire with diamond grinding wheels and to polish fused silica glass with fixed abrasive tools. As to the machining of internal cylindrical surfaces of ceramic workpieces, one-dimensional (1-D) ultrasonic vibration is applied to diamond grinding wheels [12], while in the UAFAM of hard-brittle mono-crystalline silicon and sapphire, two-dimensional (2-D), i.e., elliptic, ultrasonic vibration is superimposed onto workpieces which is distinguished from ordinary tool-vibrating methods [13,14].

Aside from grinding, the group has also originally polished fused glass by imposing elliptic ultrasonic vibration onto a fixed abrasive tool which contains softer abrasives that are able to somewhat chemically react with work-material [15, 16]. The following sections are primarily devoted to introducing the UAFAM techniques. Firstly, descriptions are made of 1-D ultrasonic assisted internal grinding (UAIG) where ultra-fine grit diamond wheels are used for grinding internal cylindrical surfaces of hard-brittle ceramics. Then, elliptic ultrasonic assisted grinding (EUAG) of hard-brittle mono-crystals is highlighted. Lastly, recently developed fixed-abrasive polishing of fused silica glass in the presence of elliptic ultrasonic vibration is narrated.

2. Ultrasonic Assisted Internal Grinding of SiC Ceramics

The rapid development of IT and medical industries requires high precision and rapid fabrication of numerous aspherical lenses. In order to satisfy the pressing requirements, hot press molding of aspherical lenses was proposed, in which molds play exceedingly important roles [17]. The molds are comprised of three types: male, female and sleeve molds (Figure 3). As the hot press molding is characterized by high temperature, the molds must stand stringent conditions. The emerging ceramics, representative of silicon carbide (SiC) exhibits potentials for such application, which is contributed to excellent thermal shock resistance and outstanding chemical stability at high temperature [18].

However, it is paradoxical that the only superb characteristics render SiC ceramics notoriously difficult to be machined. To utilize harder diamond abrasives is a common way to machine SiC [19]. Diamond grinding wheels are certainly employed to machine internal surface of SiC ceramic molds. The internal cylindrical surface should be of high accuracy and quality so that the molds will endure long-term use without significant loss of accuracy. Some disadvantages accompany conventional internal grinding (CIG) [20]: prohibitively expensive ultra-high speed grinding spindle; heating and vibration of bearings; heavy loading of grinding wheel.

Ultrasonic assisted grinding (UAG) is probably a promising improvement of conventional grinding (CG) for grinding ceramics components because the grinding efficiency and quality can be improved due to grinding force reduction and wheel loading alleviation [13]. Ultrasonic assisted internal grinding (UAIG) of ceramics were performed with metal bonded diamond wheels and it is found that grinding characteristics in UAIG, e.g. grinding forces, surface roughness, roundness and cylindricity are improved significantly compared to the CIG [21].

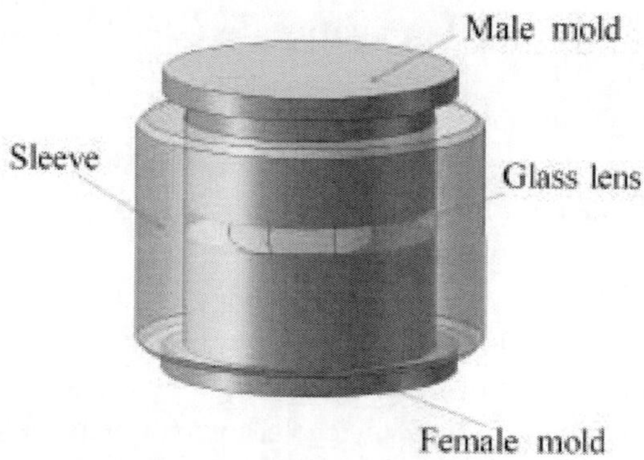

Figure 3. Structure of the mold used in hot press molding of glass lens.

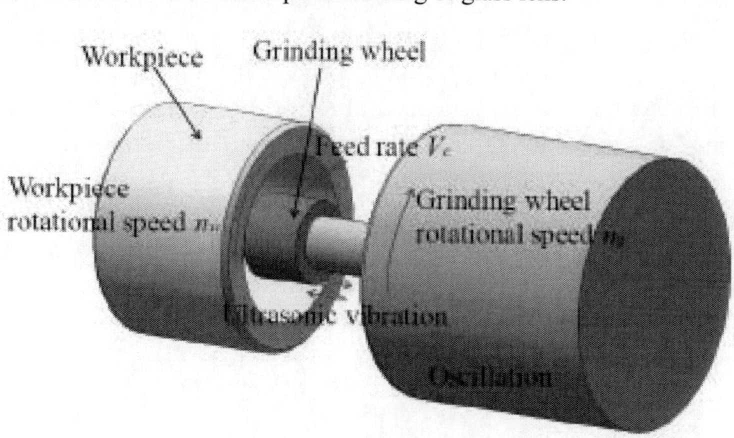

Figure 4. Schematic illustration of processing principle of UAIG.

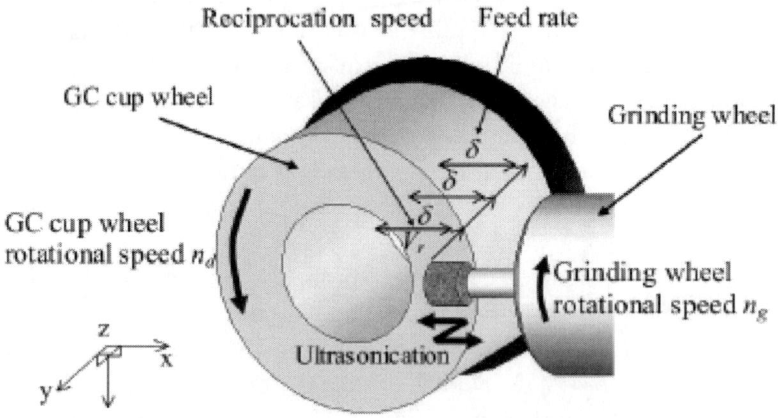

Figure 5. Schematic illustration of the ultrasonic assisted truing and dressing.

2.1. Processing Principles and Apparatus

Figure 4 schematically illustrates the processing principle of UAIG. The grinding wheel rotating at a rotational speed of n_g and ultrasonically vibrating (frequency f, amplitude A) along its axis is fed in its radial direction at a feed rate of V_c, toward the internal surface of the workpiece rotating at a rotational speed of n_w.

In addition, the grinding wheel is oscillated along its own axis frequently to machine a larger surface area and to further improve the surface quality.

A number of evidences show that the grinding accuracy is strongly related to truing/dressing accuracy of the grinding wheel. In conventional grinding, the truing/dressing accuracy is hindered when using a rotary GC cup wheel dresser because of the elastic deformation of the cantilever grinding wheel shaft caused by the truing force [22].

Hence ultrasonic assisted truing/dressing was proposed for the purpose of ameliorating truing/dressing of the grinding wheel (Figure 5). GC cup wheel rotating at n_d, reciprocates along the grinding wheel axis (x-direction) at a speed of V_r, and is fed toward the grinding wheel in its radial direction (y-axis) at a feed rate of δ per reciprocation cycle. Ultrasonic vibration (frequency f, amplitude A) is superimposed onto the x-axis and grinding wheel rotating with the x-axis at a speed of n_g.

An apparatus for implementing the UAIG is constructed by installing a commercially available ultrasonic spindle, of which a small grinding wheel is screwed onto the end face, on a commercially available NC internal grinder attached with a GC cup wheel dresser.

In addition, a dynamometer is inserted between the spindle and the worktable (Figure 6).

2.2. Grinding Characteristics

The performance of UAIG was evaluated by investigating the surface roughness, roundness and cylindricity of finished workpiece, chip formation dynamics, and grinding forces as well as grinding wheel topography. The grinding conditions are tabulated in Table 1.

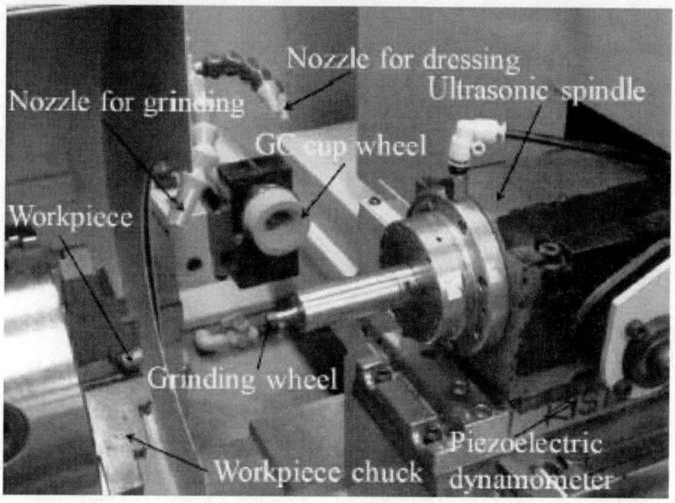

Figure 6. Experimental setup for ultrasonic assisted internal grinding.

Table 1. Processing parameters

Ultrasonic Vibration	Frequency f=40 kHz; Amplitude A=0, 4.0μm	
Grinding wheel	SD3000P100M, SD400P100M φ6×t5 mm	
	Oscillation motion	Stroke: 8mm
		Speed: 0.1m/min
	Rotational speed n_g	4000rpm,
	Feed rate V_c	10μm/min
	Wheel depth of cut	20μm
Workpiece	SiC ceramics φ12(inner)×φ22(outer)×t13 mm	
	Rotational speed n_w	300 rpm
Coolant	Solution type, 1.6%dilution	

2.2.1. Grinding Force

Grinding forces in CIG and UAIG are plotted in Figure 7. The normal and tangential grinding forces, F_n and F_t, in UAIG are decreased by 41.2% and 37.5%, respectively, compared with CIG. This decrease may be attributed to the formation of shorter and thinner chips in the presence of ultrasonic vibration [23]. In addition, the grinding force ratio of F_n/F_t was also investigated as shown in Figure 8. The value of F_n/F_t in UAIG is smaller than that in CIG, implying that the grindability of SiC ceramics is improved for the sake of the presence of ultrasonic vibration [24].

2.2.2. Roughness and Topographic Features of Workpiece Surface

Figsures 9 (a) and (b) show the photographs of SiC workpieces finished by CIG and CIAG, respectively. The surface finished by UAIG approximates a mirror surface with roughness of Ra22nm contrasting to Ra37nm in CIG. The topologies of work-surfaces by CIG and UAIG were observed with SEM as shown in Figure 10 and Figure 11, respectively. Grinding marks on the work-surface by CIG (Figure 10a) are easily discernible along the cutting direction whereas on the work-surface by UAIG (Figure 11a) the grinding marks are hardly distinguishable.

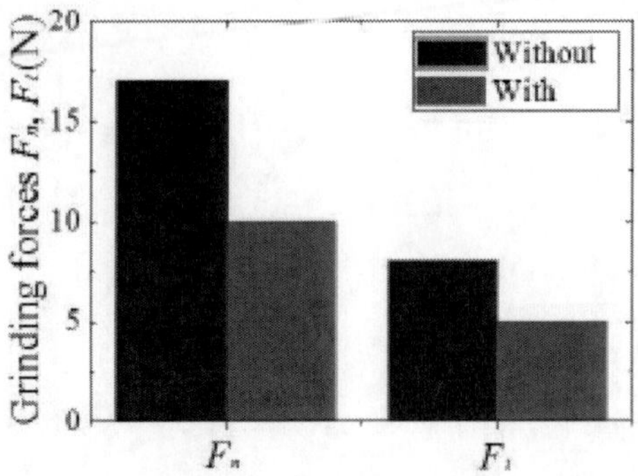

Figure 7. Grinding forces in CIG and UAIG.

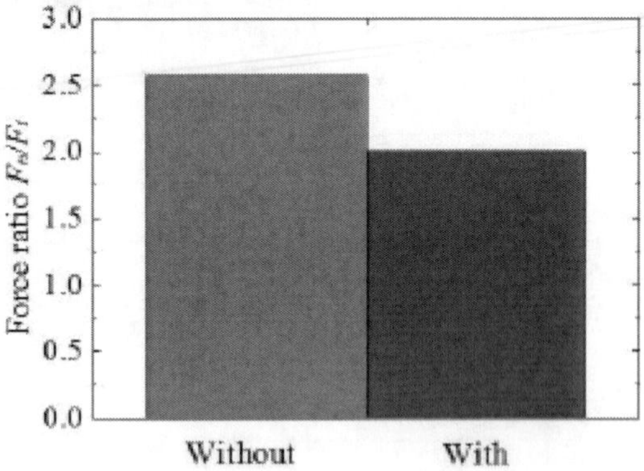

Figure 8. Grinding force ratio in CIG and UAIG.

[a]By CIG (Ra37nm).
[b]By UAIG (Ra22nm).

Figure 9. Photographs of workpieces finished by CIG and UAIG.

The reason may be that the abrasive cutting trace is generated not only along grinding speed direction but also in the wheel axial direction, resulting in the overlapping of the cutting traces by active abrasive grain cutting edges in UAIG. In addition, many brittle cracks are observed on the work-surface by CIG (Figure 10b), but the number of brittle cracks by UAIG is insignificant (Figure 11b). This is probably because the normal grinding force in the UAIG is significantly low compared with that in the CIG, resulting in few brittle cracks [25].

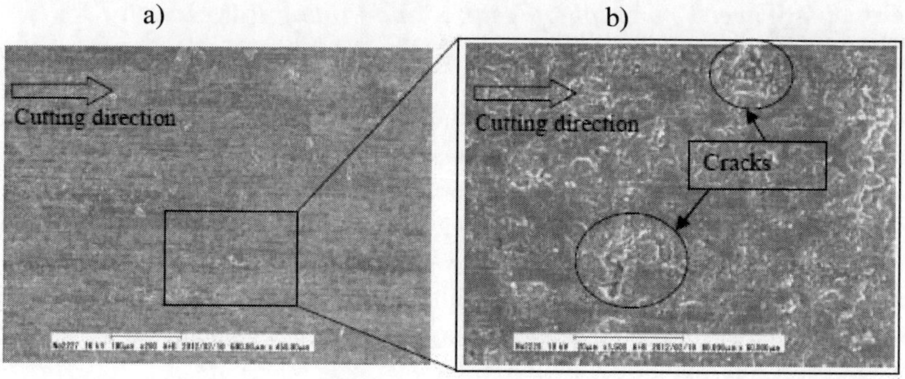

Figure 10. SEM images of workpiece surface by CIG.

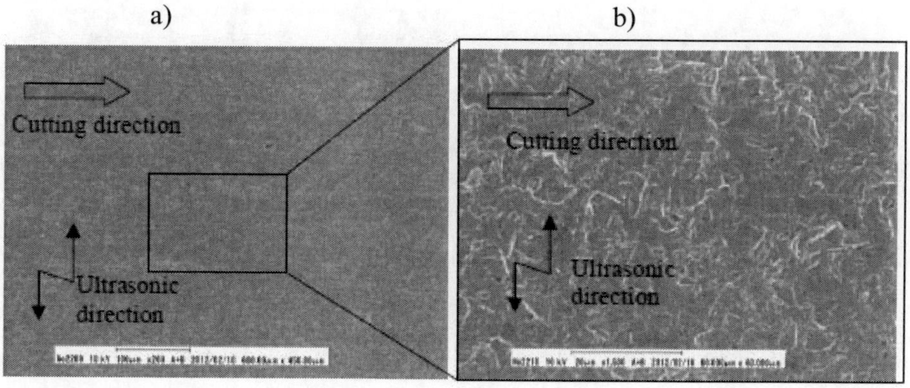

Figure 11. SEM images of workpiece surface by UAIG.

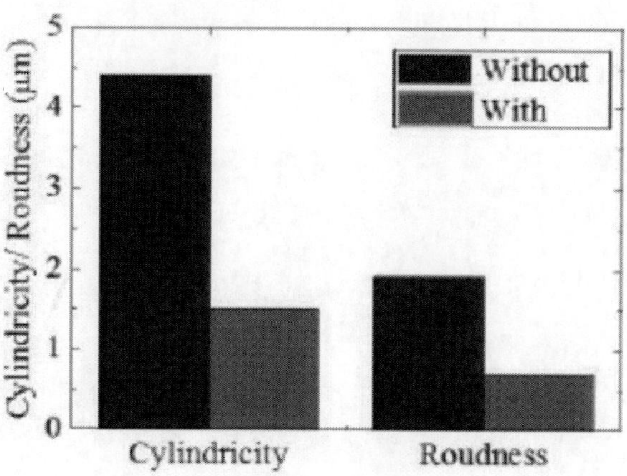

Figure 12. Cylindricity and roundness obtained in CIG and UAIG.

2.2.3. Form Accuracy: Roundness and Cylindricity

Cylindricity and roundness of the internal surface of SiC workpiece by CIG and UAIG are 4.4μm and 1.5μm, and 2μm and 0.8μm, respectively, 50% improvement by UAIG, which is due probably to the decrease in the grinding forces and the reduction of grinding force ratio in the presence of ultrasonic vibration (Figure 12).

2.2.4. Grinding Wheel Topographic Features and Cutting Edge Density

Figures 13(a) and (b) present the SEM topographies of the grinding wheel after truing/dressing and after CIG, respectively. Most cutting edges disappear after the CIG grinding even if abundant cutting edges are created after dressing/truing. This suggests that the active abrasive grains dropped out dramatically during CIG operation. By contrast, most of the cutting edges remain on the wheel working surface well even after UAIG operation (Figure 14b). The densities of cutting edges on the wheel working surfaces after truing/dressing and after CIG or UAIG were compared in Figure 15. In both CIG and UAIG, the cutting edge density decreased dramatically after grinding, although the density drops much slower in UAIG. The cutting edge density helps to achieve high quality surface, which may account for the mirror surface by UAIG.

a) b)

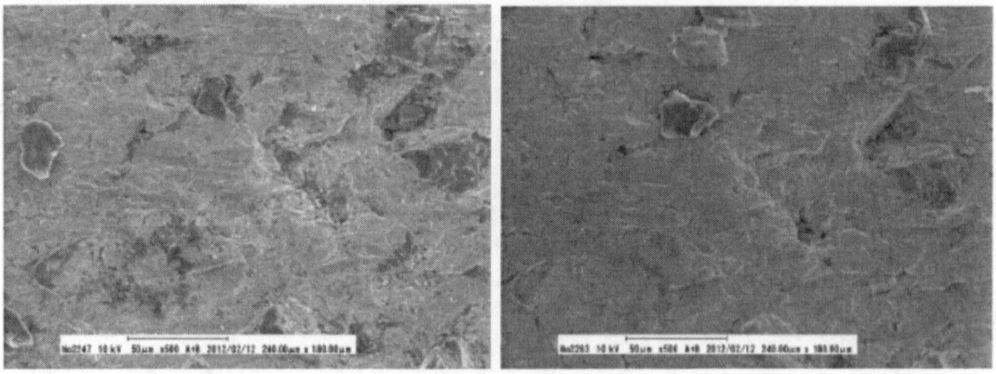

[a]After truing/dressing.
[b]After CIG.

Figure 13. SEM imagies showing the grinding wheel topographic features (a) after truing/dressing and (b) after CIG.

a) b)

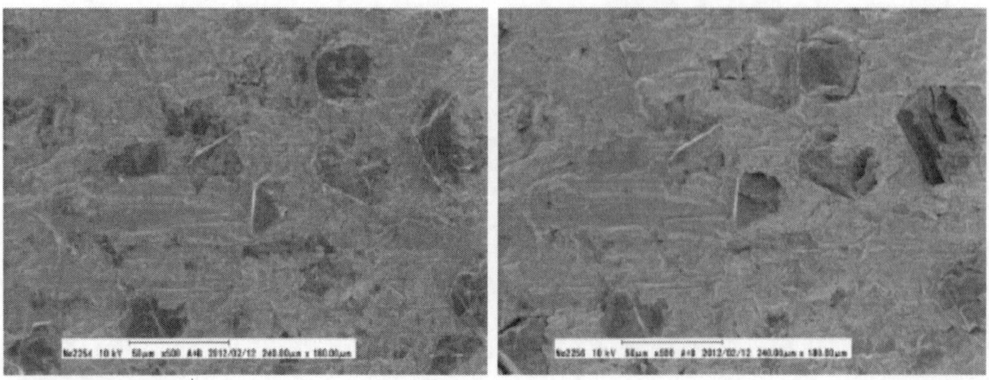

[a]After tuing/dressing.
[b]After UAIG.

Figure 14. SEM images showing the grinding wheel topographic features (a) after truing/dressing and (b) after UAIG.

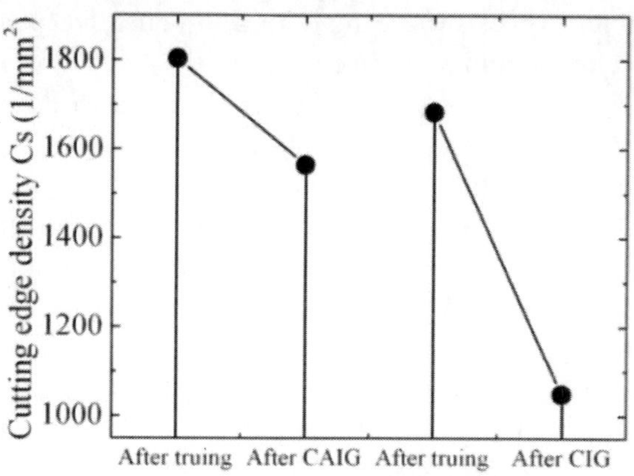

Figure 15. Decreases in abrasive grain cutting edge density after CIG and UAIG operations.

a) b)

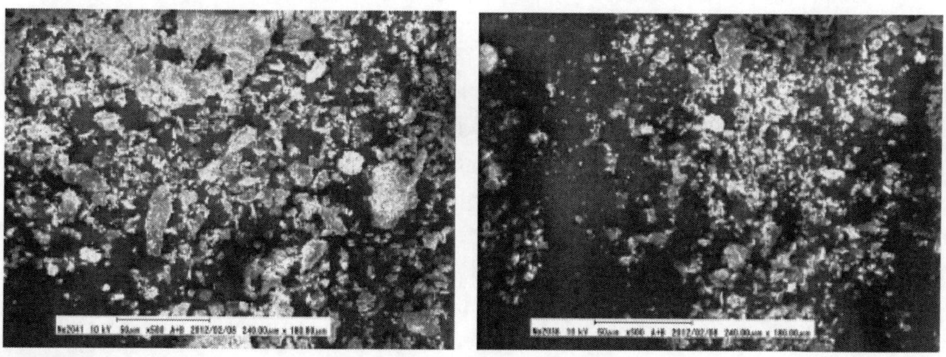

[a]Chips in CIG.
[b]Chips in UAIG.

Figure 16. SEM observation results of cutting chips generated in (a) CIG and (b) UAIG.

2.2.5. Chip Formation

Grinding chips were collected and observed using SEM. It can be found that chips generated in CIG are bigger than those in UAIG. In the other word, the chips will become thinner and tinier in UAIG (Figure 16).

3. ELLIPTIC ULTRASONIC ASSISTED GRINDING (EUAG) OF HARD-BRITTLE MONO-CRYSTALLINE MATERIALS

Hard-brittle mono-crystalline materials, e.g., sapphire, silicon carbide (SiC), silicon, have been essential in industrial applications. However, extreme hardness and brittleness cause it difficult to machine. The machining of mono-crystalline materials is in general accomplished by grinding followed by chemical mechanical polishing (CMP). Although fixed abrasive process shows potentials for replacing loose abrasive process, technical problems remain. For example, it is difficult to remove the amorphous surface layer and micro-cracks generated during grinding in a subsequent finishing operation. In particular, rapid blunting and wear of diamond abrasives on the wheel surface may cause frequent reconditioning and therefore shorten serviceable time of the grinding wheel.

Ultrasonic assisted grinding (UAG) has been introduced in different modes applied in order to develop high surface quality grinding technology. 1-D UAG like UAIG aforementioned in previous section is extensively adopted in which ultrasonic vibration applied to grinding wheels or workpiece is generated in the direction either parallel (i.e., along the wheel axis; hereafter abbreviated to AUAG) or vertical (abbreviated to VUAG) to the work-surface. Denkena *et al.* found that high surface qualities can be achieved by AUAG compared with conventional grinding without ultrasonic assistance (CG) [26]. Comparably, Mult *et al.* described the characteristics in VUAG that ultrasonic assistance leads to a significant reduction in grinding forces and an increase in MRR but a slight increase in wheel wear and surface roughness [27]. According to Mult *et al.*, the wheel wear by increased micro-fractures is, to some extent, traded off by smaller portion of grit pullout and lower thermal loading; the slight increase in roughness can be presumably ascribed to higher grain projection and less plastic deformation of the work-surface [27]. Based on 1-D UAG, Suzuki *et al.* proposed 2-D UAG that takes advantage of special grinding wheels that ultrasonically vibrates in two directions simultaneously, i.e., the tangential direction of grinding wheel and the direction vertical to the work-surface [28]. The 8-fold increase in MRR relative to constant pressure CG and up to 50% reduction of normal grinding forces in constant depth of cut grinding by this technique were experimentally evidenced, while the surface roughness shows a trivial difference from that obtained with CG [29].

Yan *et al.* performed another 2-D UAG for nano-ZrO_2 ceramics where the ultrasonic vibration applied to the workpiece is generated simultaneously in two directions, namely the tangential direction and the axial direction of grinding wheel. Better surface quality and easier realization of ductile regime grinding with the technique are reported [30].

A new 2-D, i.e., elliptic, UAG (EUAG) aimed at achieving high MRR and surface quality has been proposed by our group to machine mono-crystalline materials [13-14, 31-35]. With this method, the workpiece vibrates ultrasonically in two directions, along the grinding wheel axis and vertical to the work-surface. Based on in-house designed gadget for EUAG, a series of grinding experiments were carried out to investigate the characteristics in EUAG of mono-crystalline materials.

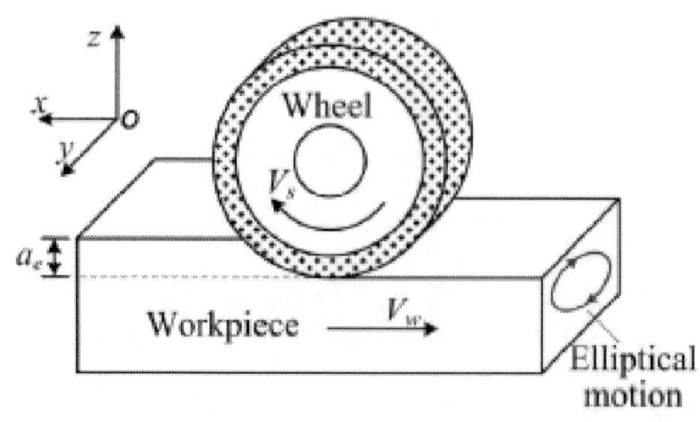

Figure 17. Processing principle of EUAG.

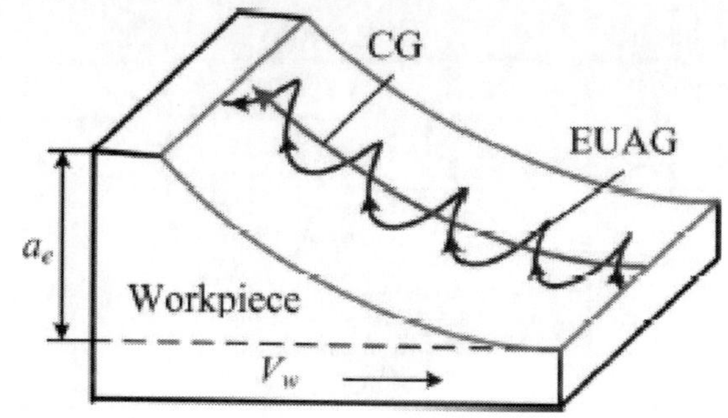

Figure 18. Grain path relative to workpiece.

a)

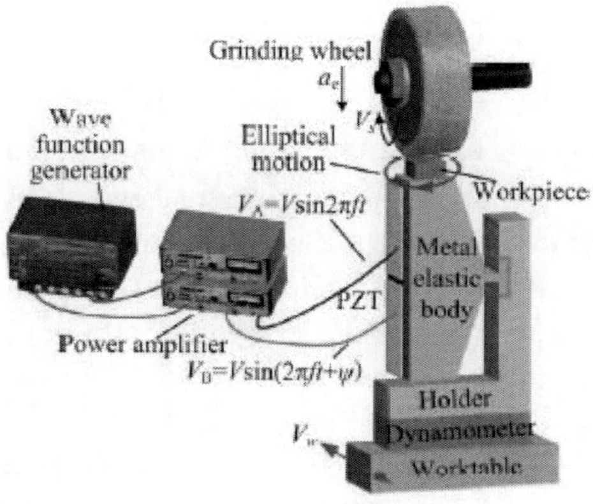

b)

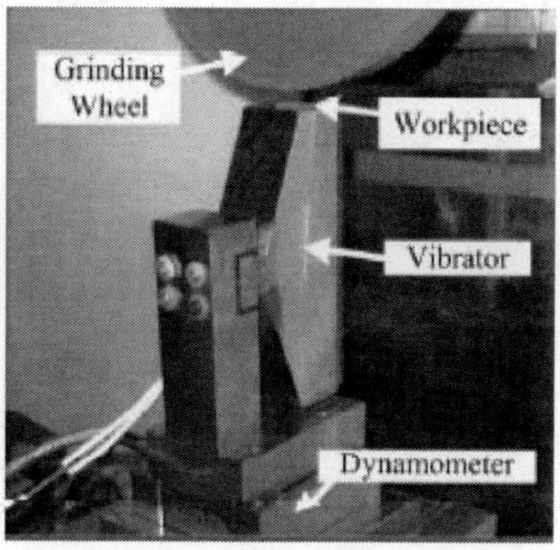

[a]Schematic illustration.
[b]Photograph of the established apparatus.

Figure 19. Experimental apparatus of EUAG.

3.1. Processing Principle and Apparatus

Ultrasonic vibrations in two directions are generated simultaneously; one is vertical to the work-surface and another along the wheel axis. The synthesis of the two vibrations results in an elliptic ultrasonic vibration which is imposed to the workpiece (Figure 17). During the EUAG, grain path relative to workpiece is a spiral curve as in Figure 18, whereas a simple arc curve is formed in CG. The horizontal vibration along the wheel axis leads to interference among grain paths. On the other hand, the vertical vibration results in periodic tool-workpiece contact, the continuously changing depth of cut, and higher engagement depths, which contribute to the significant decrease in friction, a reduction of grinding forces, and consequent increase in MRR. Figures 19(a) and (b) are the schematic and photograph of the apparatus, respectively. The EUAG unit mainly composed of an ultrasonic vibrator and a holder is installed on the worktable via a dynamometer by means of a magnetic chuck. A thin sheet-shaped workpiece is attached to the top of the vibrator. The vibrator is in practice constructed by bonding a piezoelectric ceramic device (PZT) with two separated electrodes onto a metal elastic body (stainless steel, SUS304).

a) b) c)

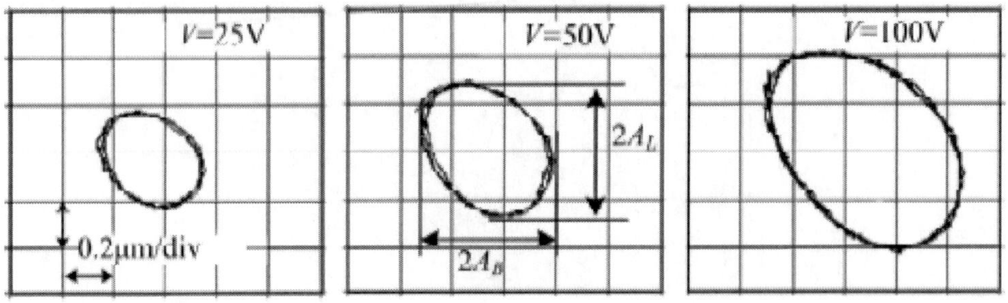

[a]25V.
[b]50V.
[c]100V.

Figure 20. Elliptical motions under different voltages at f=21.95 kHz, ψ=90°.

When two amplified alternating current (AC) signals (over 20 kHz) with a phase difference of ψ are supplied to the PZT, bending and longitudinal vibrations with amplitudes of several μm are excited simultaneously. The synthesis of vibration displacements in the two directions creates elliptical motion on the end-face of the metal elastic body.

The elliptic motions of the vibrator end-face under various phase difference were confirmed using a measuring system composed of two laser Doppler vibrometers (LV-1610 by Ono Sokki Co., Ltd.) in conjunction with a digital oscilloscope (LT364L by Iwatsu Co., Ltd.). The two radii, A_B and A_L, of the ellipse represent the amplitudes of the bending and longitudinal vibrations, respectively. The amplitude of elliptic motion increases with the increasing voltage amplitude (Figure 20).

3.2. Grinding Characteristics of Mono-Crystalline Silicon and Sapphire Materials

Grinding experiments were performed in the presence and absence of ultrasonic vibration using a resin bond diamond wheel in a CNC surface grinder (SGT-315RPA by Nagase Integrex Co., Ltd.). Thin sheet shaped mono-crystalline silicon and sapphire substrates (A-plane) with the same dimensions of 12 mm×10 mm×0.5 mm were attached onto the top of the vibrator. Diamond grinding wheels with different abrasive grain size of #400 and #800, i.e., SDC400N75BG and SDC800N75BG, were used for silicon and sapphire, respectively. The amplitude of the AC voltage applied to the PZT was varied in the range of V =25 -100V but the frequency and phase difference were kept constant at f 21.95 kHz and ψ=90°. The grinding wheel peripheral speed V_s, the work-table feed rate V_w and the wheel depth of cut a_e were V_s =10 - 30m/s, V_w =50 - 400mm/min, a_e=0.5 - 3μm. No any coolant was employed in the grinding.

3.2.1. Grinding Forces

Figure 21 shows the influence of applied voltage on grinding forces where the F_n, F_t, F_a are the normal, tangential and axial components of the grinding force, respectively. It is obvious that the grinding forces in EUAG are less great than that in CG and drop as the voltage increases. Furthermore, it is found that both the F_n and F_t in EUAG are significantly smaller than those in CG up to 40% (Figure 22). Two main reasons are postulated: (1) ultrasonic assistance augments the acceleration and velocity of abrasive grain relative to the workpiece, and thus the workpiece is impacted and interpenetrated by high frequency interaction of grains, improving the machinability of workpiece; (2) the tool-workpiece contact is interrupted periodically and the contact of the individual diamond grit is shortened, reducing frictional effects and decreasing normal grinding forces on average.

3.2.2. Chip Formation

Silicon chips by CG and EUAG are arrayed in Figure 23, showing that the elliptic vibration leads to the formation of thicker and shorter chips. It is also found that the chip by CG is smoother than those in EUAG. The greater chip thickness corresponds to lower specific energies and the average temperature on the finished surface is proportional to the specific energy [34]. From this view of point, the grinding temperature in EUAG should be lower than that in CG, which benefits thinner deformed layer and better ground surface.

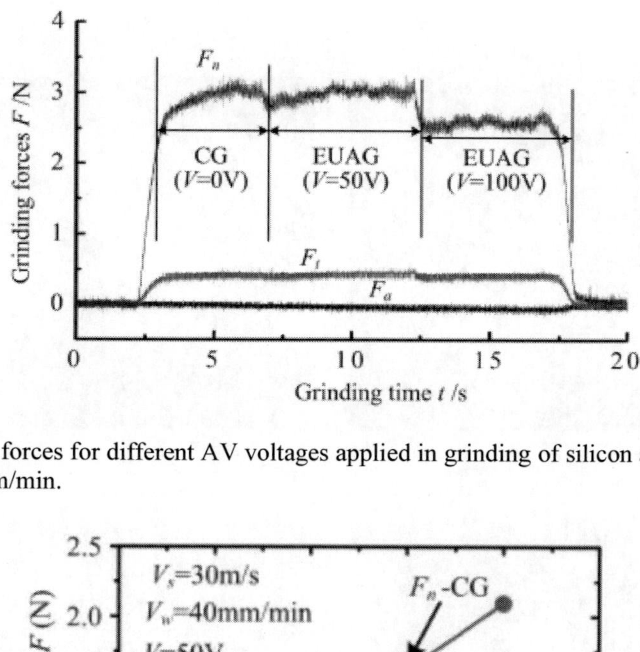

Figure 21. Grinding forces for different AV voltages applied in grinding of silicon at a_e=2 μm, V_s=20 m/s, and V_w=100 mm/min.

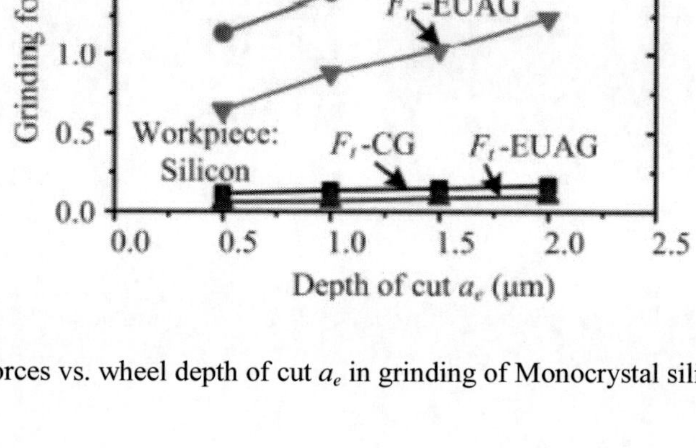

Figure 22. Grinding forces vs. wheel depth of cut a_e in grinding of Monocrystal silicon.

The ultrasonic vibration alters single-direction motion of grains and increases the maximum depth of cut of a single grain, which is conducive to the formation of a larger number of thicker and shorter chips.

3.2.3. Roughness and Integrity of Workpiece Surface

The workpiece surface roughness of monocrystal silicon and sapphire in CG and in EUAG at different depth of cuts is shown in Figures 24(a) and (b), respectively.

It can be seen that although the surface roughness increases with increasing wheel depth of cut, the EUAG decrease the surface roughness by 25%. The results can be interpreted by interference among different abrasive paths and overlapping of abrasives [31-33].

Higher surface quality can be achieved in EUAG, while the surface obtained in CG contains some deep grooves and fractures (Figures 25 and 26). The greater grinding forces in CG induce more micro-fractures and severer cracks on the ground surface.

Once the elliptic ultrasonic vibration is applied to the workpiece, the ultrasonic vibration alleviates friction effects and grinding forces and thermal loads, which decreases fractures and cracks on the ground surface.

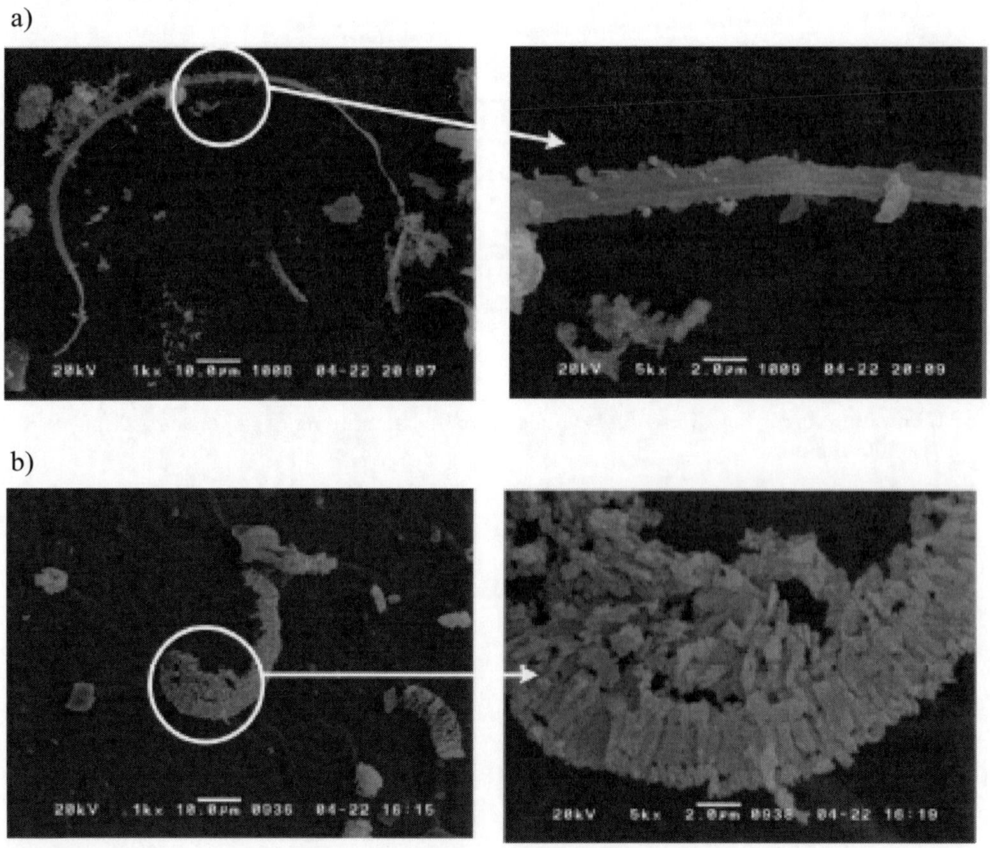

[a]CG (V=0V).
[b]EUAG (V=100V).

Figure 23. SEM images of silicon grinding chips at V_s=30 m/s, a_e =2 μm and V_w=100 mm/min.

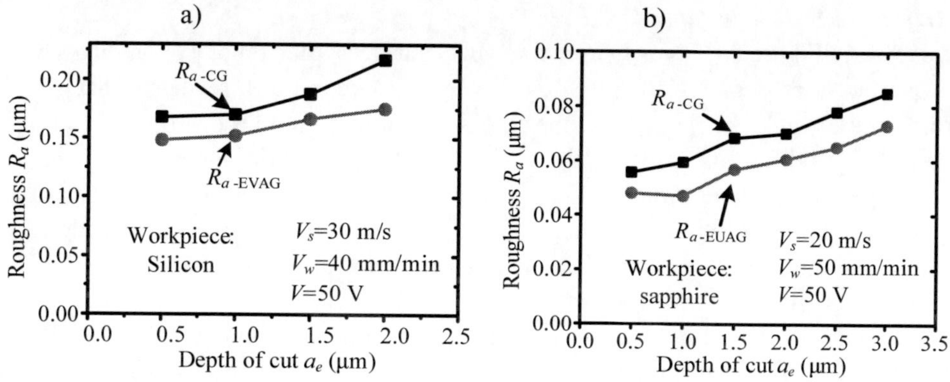

[a]Monocrystal Silicon workpiece.
[b]Sapphire (A-plane) workpiece.

Figure 24. Workpiece surface roughness vs. depth of cut.

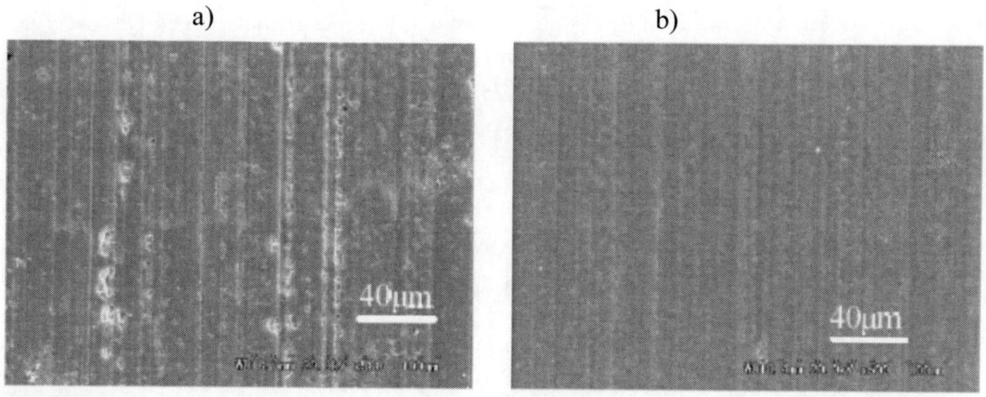

[a]CG (V=0V).
[b]EUAG (V=50V).

Figure 25. SEM images of the ground silicon surfaces at a_e =0.5 μm, V_s =30 m/s and V_w =20 mm/min.

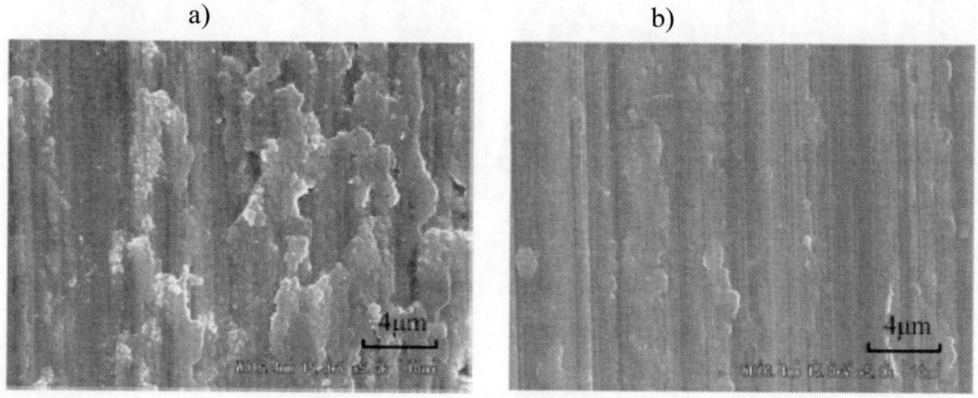

[a]CG (V=0V).
[b]EUAG (V=50V).

Figure 26. SEM images of the ground sapphire surface at a_e=2 μm, V_s=20 m/s, and V_w=50 mm/min.

3.2.4. Grinding Wheel Wear

As shown in Figure 27, investigating the wheel surfaces after sapphire grinding with a 3D laser scanning microscope (VK-8700 by Keyence Corporation) revealed that the bond fracture occurs in CG. This is because the forces acting on a grain are sufficient to break the bond, generating bulky pits and the slight wheel loading and chip adhesion onto the wheel surface. Attritious wear and grain pullouts of the grinding wheel lower the number of active cutting edges. Although the grain flattening and the decrease in active cutting edge density appear in EUAG, the wheel loading and chip adhesion was rarely found and the active cutting edge density is greater than that in CG.

The grinding wheel in EUAG retains higher grinding performance than that in CG. From the topography of abrasive grain (Figure 28), it can be seen that in CG wear of a grain is generated due to attritious wear and chips are located between grains. Due to great grinding forces, the macro-fracture of grain occurs. However, in EUAG micro-fracture and cleavage of abrasive grain take place and slight flattening occurs, indicating that the abrasive cutting edge is sustained. In addition, some micro-grooves on the grain surface are identified, which is due hypothetically to intermittent cutting of the diamond grain. As sketched in Figure 29(a), the wheel wear in CG can be characterized by attritious wear, grain macro-fracture, and grain pullout. In stark contrast, the wheel wear in EUAG can be ascribed to attritious wear, grain micro-fracture, and chip adhesion (Figure 29(b)).

a)

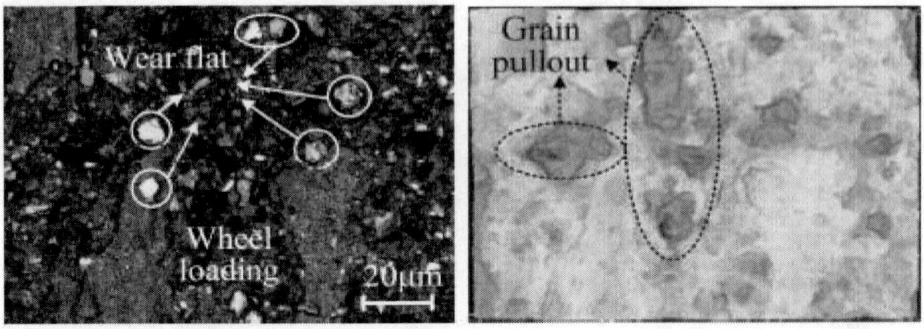

b)

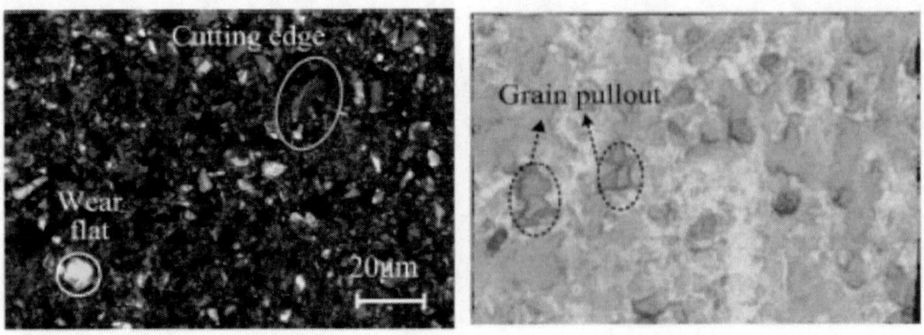

[a]CG.
[b]EUAG.

Figure 27. 2D (left side) and 3D (right side) laser microscopic images of the wheel surface after sapphire grinding for 40 passes.

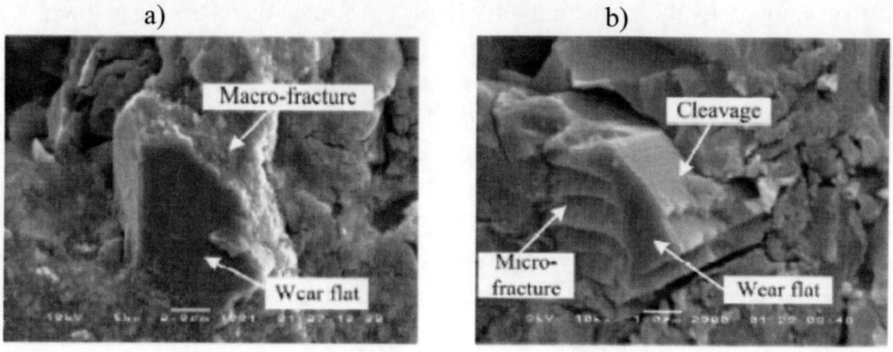

[a]CG.
[b]EUAG.

Figure 28. Topography of abrasive grain in the wheel surface after sapphire grinding 40 passes.

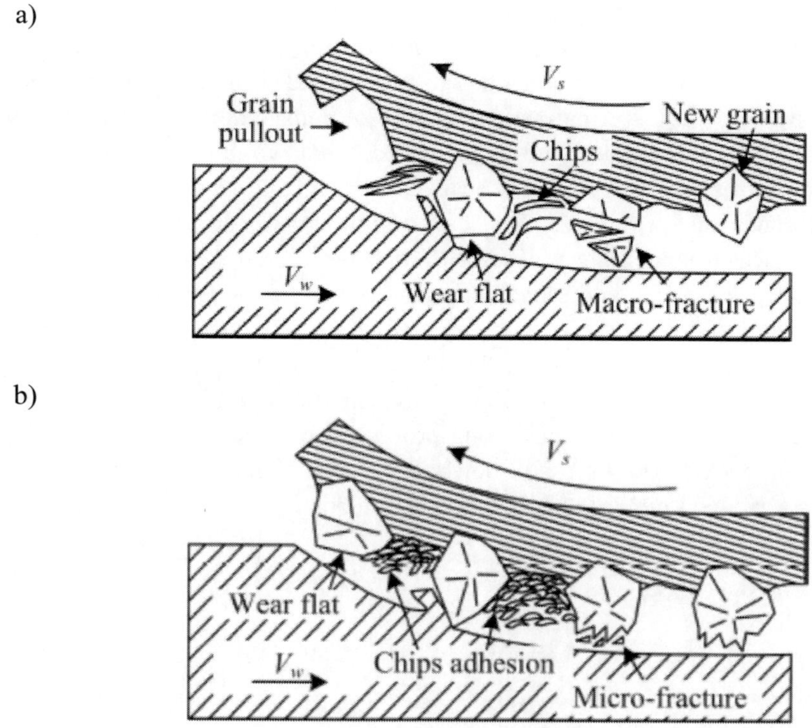

[a]CG.
[b]EUAG.

Figure 29. Mechanisms of wheel wear during sapphire grinding.

4. Ultrasonic Assisted Fixed Abrasive Polishing of Hard-Brittle Glass

Glass polishing, an ancient craft dating back to Newton's time, is usually performed by using loose abrasive. The loose abrasive polishing is still a prevailing finishing technique, which is also applied to the polishing of silicon substrate, dielectrics and metals and the planarization of shallow trench isolation (STI).

Typical loose abrasive polishing is comprised of three factors: polishing tools, polishing slurry and workpiece. The polishing is fulfilled by the interaction among tools, slurry and workpiece. The motion of abrasives is stochastic and renders the process unpredictable, resulting in non-deterministic characteristics of the process. Aside from the physical nature of the loose abrasive process, chemical reaction is considered to be of paramount importance to the polishing processes that employ abrasive softer or equivalent to the workpiece to be polished in hardness. This kind of loose abrasive polishing possesses obvious superiorities: generating ultra-smooth surface (surface roughness RMS<1.0nm), inducing little surface and subsurface damage. However, some inherent shortcomings accompany the process, such as frequent dressing of polishing pad, minor "active" abrasives (<0.5%), lack of determinism, etc. Hence some institutes and organizations have turned to fixed abrasive polishing [36-38]. Fixed abrasive polishers will be instrumental in automatizing polishing process and facilitating predictable manufacture. The abrasives are imbedded in a matrix made of polymers and no additional slurry is supplied during manufacturing. The distribution and the trajectories of the abrasives, to a certain extent, can be modeled. Moreover, the number of "active" abrasives participating in material removal is considerably increased. As a consequence, the process is more deterministic than loose abrasive polishing. As it is well known, machining performance, at least MRR, can be improved by introducing ultrasonic vibration into manufacturing processes. The easy-to-implement property adds to the universality of ultrasonics to machining processes. Ultrasonic vibration has been combined with chemical mechanical polishing and satisfactory results have been demonstrated [39-40]. The vibration is superimposed onto a special wafer carrier and the carrier together with the wafer vibrates in vertical direction (Figure 30). The smoother surface and greater MRR have been confirmed. Inspired by the fact that machining performance can be ameliorated on the introduction of ultrasonic vibration into machining processes, our group "appropriates" this creed to amalgamate ultrasonic vibration with newly developed fixed abrasive polisher with anticipation of the improvement either in MRR or in surface quality.

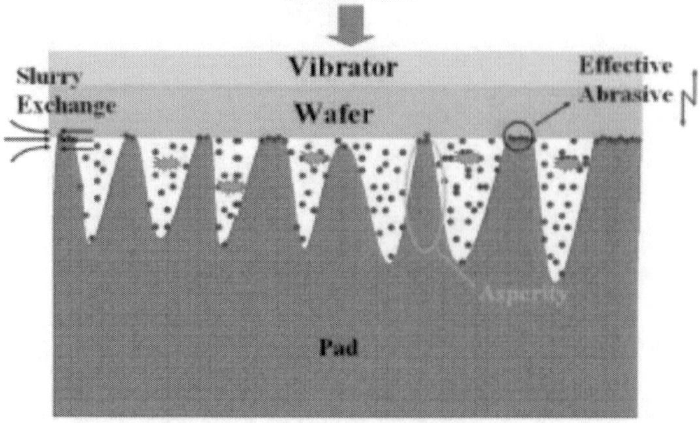

Figure 30. Ultrasonic assisted CMP (chemical mechanical polishing). The carrier holding the wafer can vibrate in the direction normal to wafer surface. The slurry supply and the movement of wafer and pad are identical to that in conventional CMP (with the permission of Inderscience Publishers).

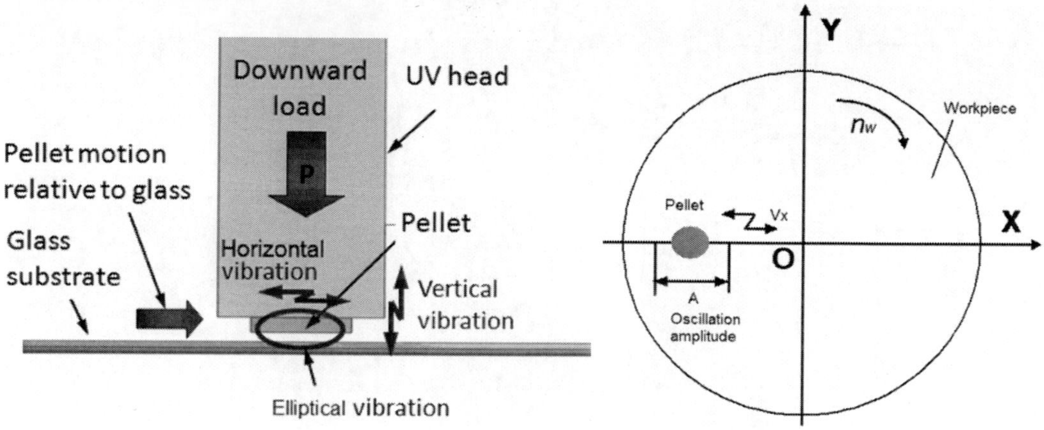

Figure 31. Illustration of processing principle of UFAP.

Our preliminary results of ultrasonic fixed-abrasive polishing (UFAP) suggest that MRR is considerably enhanced after integrating ultrasonic vibration whilst surface roughness does not deteriorate appreciably. The related experimental setup is detailed, which is followed by our latest results.

4.1. Processing Principle and Apparatus

Figure 31 schematically illustrates the processing principle of ultrasonic assisted fixed abrasive polishing (UFAP). An ultrasonic vibration (UV) head, onto which a polishing pellet is glued, is pressed against the workpiece (i.e., glass substrate) surface at a downward load of P. An elliptic ultrasonic vibration (UV) is created on the lower end face of the UV head by the simultaneous generation of two ultrasonic vibration modes, longitudinal vibration (L1 mode) vertical to the workpiece surface and bending vibration (B4 mode) parallel to the workpiece surface. Let the origin of a XYZ-coordinate system be at the workpiece center, the X and Y-axes be on the workpiece surface, and the Z-axis be vertical to the workpiece surface, the L1 and B4 modes are in Z- and X-directions, respectively, resulting in the elliptic ultrasonic vibration (L1B4 mode) of the pellet in XZ-plane. When the workpiece is rotated around Z-axis at a rotational speed of n_w, and the UV head is oscillated along X-axis at a stroke of A and a speed of V_x, a constant-pressure UFAP operation is thus performed to finish the demanded area of workpiece surface.

Based on the processing principle, a customized apparatus was created (Figure 32). The apparatus consists of a UFAP unit composed of a PZT vibrator (i.e., the UV head), a mass-spring-damper mechanism, a dynamometer and its holder, a linear motion actuator for the oscillation (i.e., cross motion) of the unit along X-axis, a mechanism for the vertical feed motion of the unit, a vacuum chuck for holding the workpiece, and a spindle for rotating the vacuum chuck. In the UFAP unit, the mass-spring-damper mechanism is installed in order to guarantee that the pellet attached on the lower end face of the PZT vibrator has flexible contact with the workpiece in constant-pressure machining. The machining pressure is controlled by adjusting the vertical feed with the hand wheel and determined quantitatively with a strain gauge dynamometer.

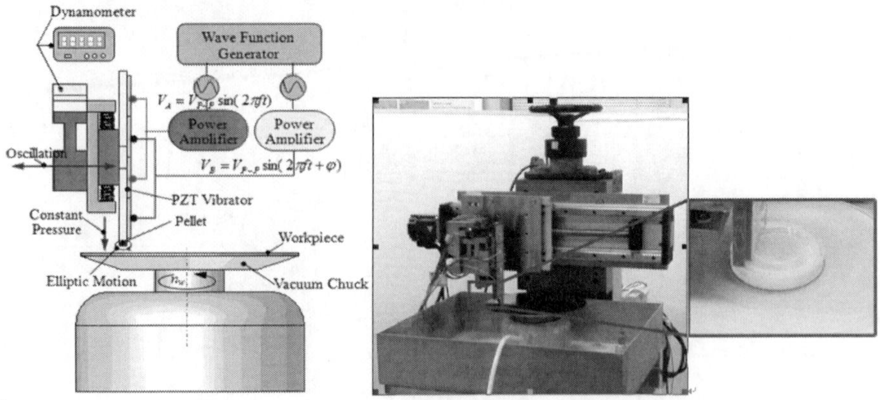

Figure 32. Schematic illustration of UFAP apparatus.

a) b) c)

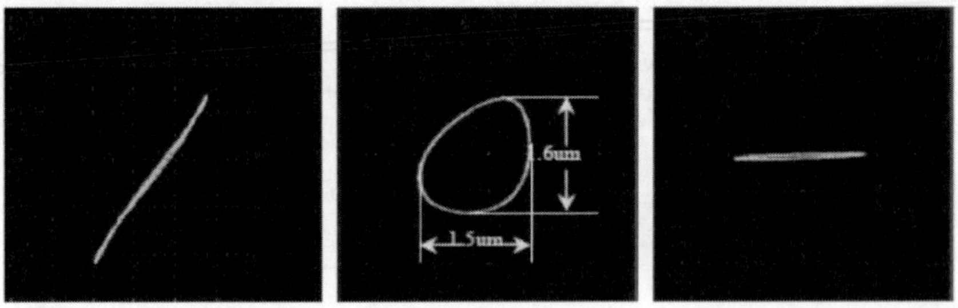

[a]0°.
[b]135°.
[c]180°.

Figure 33. Trajectories of vibrator at varied phase differences when f=15.3kHz, $V_{p\text{-}p}$=150V.

In this apparatus, the PZT vibrator plays a key role in giving the pellet an ultrasonic vibration. The vibrator has the same structure and the creation method of elliptic ultrasonic vibration is also the same as that used in section 2.1(Figure 19). Ultrasonic vibration amplitude is roughly proportional to the applied voltage $V_{p\text{-}p}$ while the vibration trajectory is affected significantly by the phase difference ϕ (Figure 33). When the phase difference is equal to 0° and 180°, the vibration is close to a straight line; otherwise, the vibration is elliptic.

4.2. Machining Characteristics of Fused Silica Glass

A circular fused silica sample with dimension of Φ50mm×t10mm and Initial surface roughness of ~200nmRa was placed on the vacuum chuck as the workpiece. The possible effects of processing parameters and specifications of ultrasonic vibration, i.e. feed rate and stroke in X-axis, downward load, rotational speed, applied voltage and phase difference, on the MRR and surface roughness were investigated systematically under the presence (denoted "with UV")/absence (denoted "without UV") of ultrasonic vibration. Table 2 shows the processing parameters.

Table 2. Processing parameters

Pellets	Dimension	Φ4mm×1mm
	Abrasive	CeO_2
Ultrasonics	Frequency	15.3kHz
	Applied voltage $V_{p\text{-}p}$	150V
	Phase difference	135°
Pellet oscillation along X-axis	Stroke A	3mm
	Speed V_x	3mm/s, 6mm/s, 9mm/s
Downward load	P	~4N, ~8N, ~12N
Workpiece rotational speed	n_w	~400rpm, ~600rpm, ~800rpm

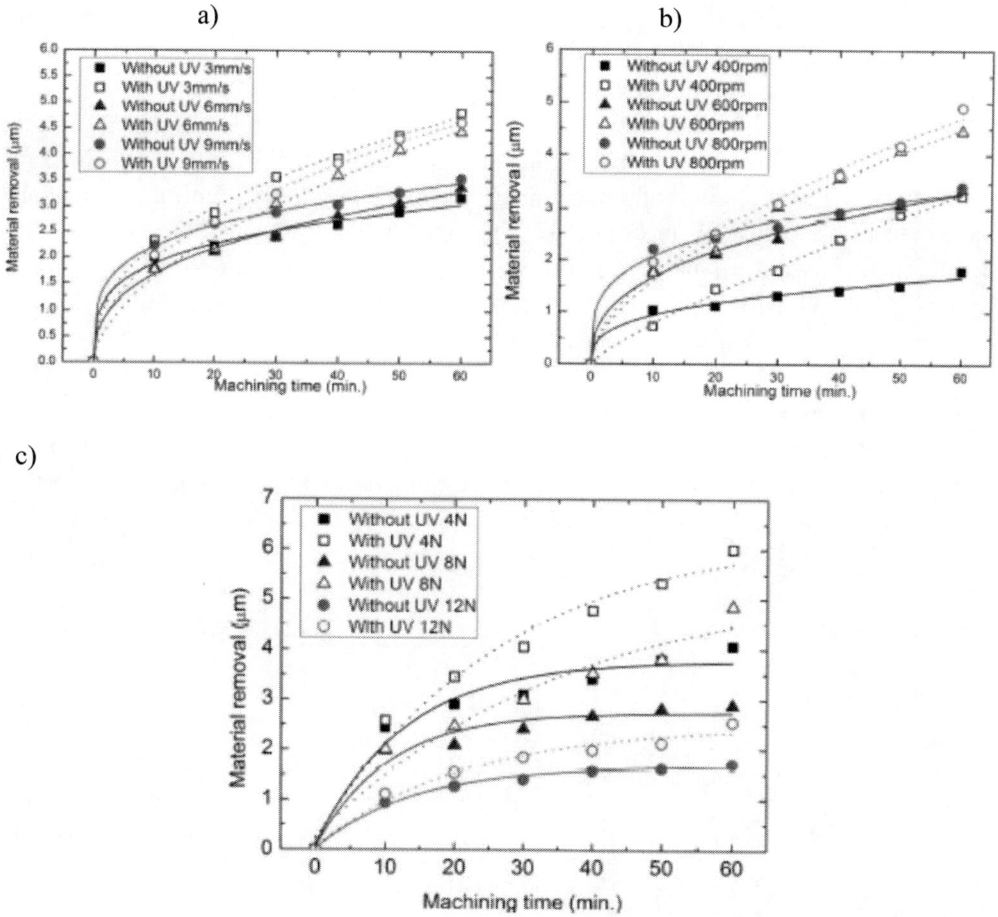

Figure 34. The influence of processing parameters on material removal rate (a) oscillation speed V_x at n_w=600rpm and P=4N, (b) rotational speed n_w at V_x=6mm/s and P=4N, (c) downward load P at V_x=6mm/s, n_w=600rpm, in "without UV" and "with UV" polishing.

4.2.1. Material Removal Rate

Referring to Figure 34, it is understood that the MRR is indeed increased by introducing ultrasonic vibration, regardless of the processing parameters and is marginally influenced by the oscillation speed.

It is noteworthy that increasing the downward load paradoxically decreases the MRR for the experimented loads, which ostensibly contradicts common creeds held by polishing community that the MRR should ascend with the elevated pressure. The results on UV specifications demonstrate that the MRR is the greatest when the vibration motion is elliptical and increases with the applied voltages under experimented conditions (Figure 35). The excellent performance of elliptic vibration is attributed to dispelling of polishing debris at the interface of abrasive and workpiece. Increasing the applied voltage and resultant vibration amplitude will also enhance the dispelling of debris.

The debris is comprised of ceria and silica and the silica will undermine the chemical reaction between ceria abrasives and glass workpiece.

4.2.2. Roughness and Topographic Features of Workpiece Surface

Another phenomenon can also be readily identified that the ultrasonic vibration mildly deteriorates the surface roughness in the present work. The surface roughness was also compared after there were no discernible cracks by grinding. The results show that the surface roughness Ra by "with UV" polishing populates the range of 2~5nm while that by "without UV" mostly falls into 1~4nm, lightly smaller than that in "with UV" polishing (Figure 36). The best surface roughness by both machining techniques is <2.0nm.Typical morphology of machined surface is displayed in Figure 37. The fused silica by "with UV" process takes a patterned surface with a spatial period of ~35μm. The computed period is ~41μm ($=2\pi n_w rt=\sim 41\mu m$) consistent with the measured one, implying that the patterned surface results from ultrasonic vibration.

4.2.3. Material Removal Mechanism

Carefully comparing the surface of polishing pellets subjected to different polishing pressure, it can be recognized without much difficulty that certain foreign/new substances was produced.

a) b)

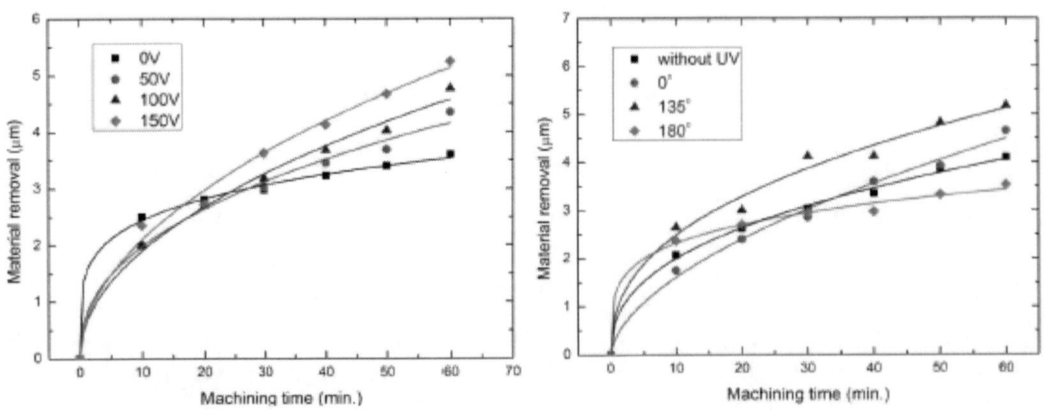

Figure 35. The material removal due to varied input voltages and phase differences. The greater the input voltage is, the greater the material removal is. The greatest material removal is found under the conditions of elliptical vibration. The reason is reckoned to be due presumably to exceptional ability of elliptic vibration to dispel polishing debris that hampers the chemical actions between abrasives and glass.

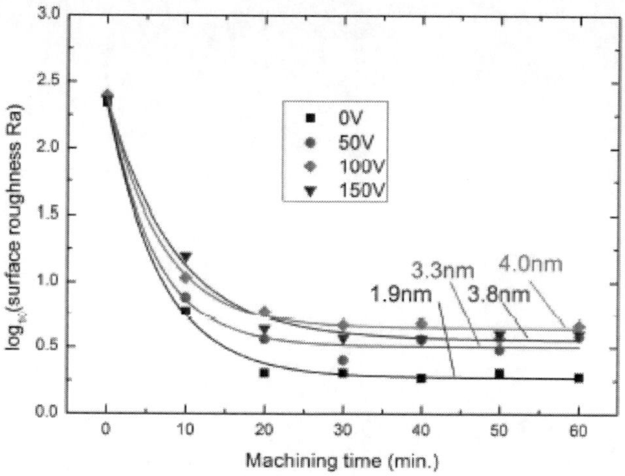

Figure 36. Variation of surface roughness during polishing.

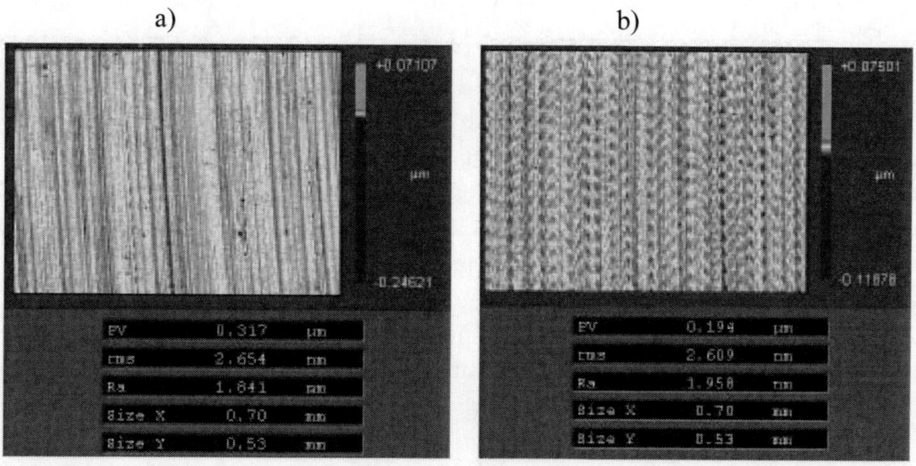

Figure 37. Surface morphologies machined by (a) "without UV" and (b) "with UV".

The new substances are suspected to be directly responsible for the declined MRR because the material is unlikely to be removed purely mechanically by CeO_2 abrasives since the hardness of ceria is of lower hardness than fused silica. It is chemical actions that enable the removal of material from the glass surface in our cases.

The material removal process is conceived as follows (Figure 38): firstly, the ceria reacts with silica in solid-state phase under the circumstances of exceedingly high pressure to form new substances with lower hardness than glass bulk and ceria on the topmost of fused silica and then the resultant softer substances are removed by ceria abrasive mechanically and plastically; alternatively, the Ce in ceria abrasives bonds with Si in glass, next silica material is torn away from glass as a lump on account of greater strength of Ce-O bond than Si-O bond and lastly the lump is disengaged and this way the glass material is removed. The actual mechanism of material removal needs further investigations. Returning to the point, the reduction in MRR is ascribed to the adsorption of silica onto the surface of ceria abrasive as it could hinder and even stop chemical reactions between ceria and silica.

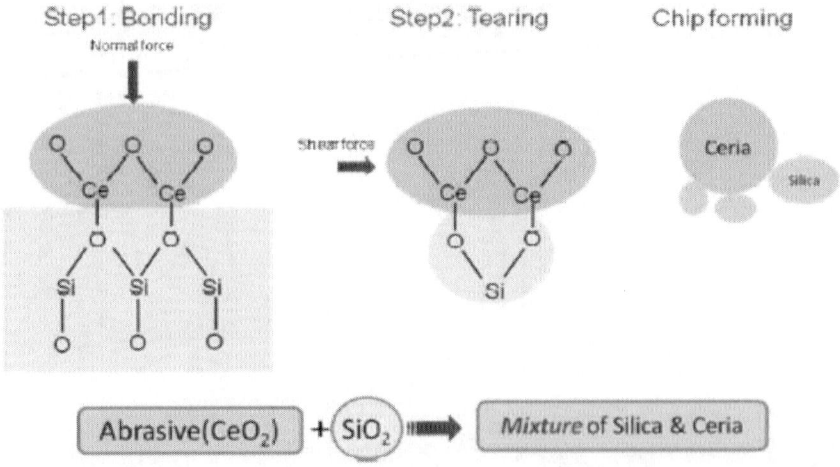

Figure 38. Proposed mechanism in fixed abrasive polishing of glass. The ceria bonds with silica because of extreme pressure and external shear force will break the Si-O bond to induce material removal. Lastly, the silica will adsorb onto the ceria surface to constitute debris.

As a consequence, the MRR is decreased. In our experiments, less intense pressure is preferable so that the polishing can proceed smoothly. The reason for the outstanding performance in MRR by "with UV" process may lie in the exceptional capacity to dispose the polishing debris between pellets and surface being machined. The debris is considered to be detrimental to machining in grinding process. In our experiments, the debris resulting from the chemical reaction products and/or silica can increasingly cover the surface of pellets and thereby undermine potential reactions between CeO_2 and SiO_2. As a result, the MRR is decreased accordingly because of the accumulation of polishing debris. The material is removed due crucially to the synergy of chemical and mechanical actions. It is presupposed that ceria bonds with silica to form Ce-O-Si systems and then the Si-O de-bonds owing to the greater strength of the Ce-O. This way, the glass is polished and debris forms. If the debris caulks at the interface of the polisher and glass workpiece, the MRR will drop because silica on the surface of debris suppresses the chemical bonding, which also corroborates indirectly the interpretation for greater MRR in UV-assisted polishing.

CONCLUSION

Our group has endeavored on three kinds of UAFAM (ultrasonic assisted fixed abrasive machining) methods, UAIG (ultrasonic assisted internal grinding), EUAG (elliptic ultrasonic assisted grinding) and UFAP (ultrasonic assisted fixed abrasive polishing), in anticipation of providing alternative and novel machining techniques for the precision and efficient machining of hard-brittle materials.

These techniques differ from each other in processing principles and performances.

1 In the UAIG of SiC ceramics, grinding forces are decreased by several tens percentage once the UV is applied to the grinding wheel. The ratio of the normal grinding force to the tangential grinding force can also be decreased by the

application of UV, indicating the grindability of SiC ceramics is improved. Workpiece surface quality can be significantly improved with UAIG: surface roughness is decreased by ~40% and fewer brittle cracks are generated on the workpiece surface. The cylindricity and roundness of workpiece in UAIG is smaller by around 50% than those in CIG. In UAIG, the workable conditions of wheel surface lasts longer than CIG.

2 In the EUAG, the workpiece instead of grinding wheel vibrates elliptically. As compared to CG, the EUAG: can achieve better workpiece surface quality; is capable of increasing the MRR; can give rise to thicker and shorter grinding chips, which is beneficial to grinding wheel loading and reserving the workability of grinding wheel.

3 The MRR of UAFP is raised compared to fixed abrasive polishing without UV. Increasing the UV amplitude, the MRR is raised correspondingly. It is the outstanding debris-dispelling ability of UV-assisted polishing that boosts the MRR in UAFP. The material is removed because of the synergy of chemical and mechanical actions in fixed abrasive polishing with and without ultrasonics.

REFERENCES

[1] Markov, *Ultrasonic Machining of Intractable Materials*, (Mashgiz, Moscow, USS.R.,) 1962 (in Russian) [translated by Scripta Technica Ltd., edited by E. A. Neppiras, (Iliffe Books Ltd., London, UK), Chap. 2, 1966].

[2] P. L. Guzzo, A. H. Shinohara and A. A. Raslan, "A comparative study on ultrasonic mchining of hard and brittle materials", *J. Braz. Soc. Mech. Sci. Eng.* 26, 56-61, 2004.

[3] T. B. Thoe, D. K. Aspinwall and M. L. H. Wise, "Review on ultrasonic machining", *Int. J. Mach. Tool. Manuf.* 38, 239-255, 1998.

[4] Y. Ichida, R. Sato, Y. Morimoto, K. Kobayashi, "Material removal mechanisms in non-contact ultrasonic abrasive machining", *Wear* 258, 2005, 107-114, 2005.

[5] D. E. Brehl and T. A. Dow, "Review of vibration-assisted machining", *Prec. Eng.* 32, 153-172, 2008.

[6] T. Moriwaki, E. Shamoto and K. Inoue, "Ultraprecision ductile cutting of glass by applying ultrasonic vibration", *Ann. CIRP* 41, 141-144, 1992.

[7] M. Zhou, X. J. Wang, B. K. A. Ngoi and J. G. K. Gan, "Brittle-ductile transition in the diamond cutting of glasses with the aid of ultrasonic vibration", *J. Mater. Process. Technol.* 121, 243-251, 2002.

[8] W. Qu, K. Wang, M. H. Miller, Y. Huang and A. Chandra, "Using vibration-assisted grinding to reduce subsurface damage", *Prec. Eng.* 24, 329–337, 2000.

[9] C. Nath and M. Rahman, "Effect of machining parameters in ultrasonic vibration cutting", *Int. J. Mach. Tool. Manuf.* 48, 965-974, 2008.

[10] M. Zhou, B. K. A. Ngoi, M. N. Yusoff and X. J. Wang, "Tool wear and surface finish in diamond cutting of optical glass", *J. Mater. Process. Technol.* 174, 29-33, 2006.

[11] K. Egashira and T. Masuzawa, "Microultrasonic machining by the application of workpiece vibration", *Ann. CIRP* 48, 131-134, 1999.

[12] J. Cao, Y. Wu, H. Guo, Y. Li, M. Fujimoto and A. Ohmura, "Simulation investigation of chip formation in ultrasonically assisted grinding of SiC ceramics with single diamond abrasive grain", *Proc. of ICPE2012*, to be published.

[13] Z. Liang, Y. Wu, X. Wang, W. Zhao, "A new two-dimensional ultrasonic assisted grinding (2D-UAG) method and its fundamental performance in monocrystal silicon machining", *Int. J. Mach. Tools Manuf.* 50, 728-736, 2010.

[14] Z. Liang, X. Wang, Y. Wu, L. Xie, Z. Liu, W. Zhao, "An Investigation on Wear Mechanism of Resin-bonded Diamond Wheel in Elliptical Ultrasonic Assisted Grinding (EUAG) of Monocrystal Sapphire", *J. Mater. Proc. Technoll.* 212, 868-876, 2012.

[15] Y. Li, Y. Wu, J. Wang, W. Yang, Y. Guo and Q. Xu, "Tentative investigation towards precision polishing of optical components with ultrasonically vibrating bound-abrasive pellets", *Opt. Express* 20, 568-575, 2012.

[16] Y. Li, Y. Wu, L. Zhou, H. Guo, J. Cao, "Investigation into Chemo-Mechanical Fixed Abrasive Polishing of Fused Silica with the Assistance of Ultrasonic Vibration", *Proc. of ICPE2012,* to be published.

[17] Y. Namba, T. Shimomura, A. Fushiki, A. Beaucamp, I. Inasaki, H. Kunieda, Y. Ogasaka and K. Yamashita, "Ultra-precision Polishing of Electroless Nickel Molding Dies for Shorter Wavelength Applications", *Ann. CIRP* 57, 337-340, 2008.

[18] C. Hall, M. Tricard, H. Murakoshi, Y. Yamamoto, K. Kuriyama and H. Yoko, "New mold manufacturing techniques", *Proc. SPIE* 5868, 58680V, 2005.

[19] S. Agarwal, P. V. Rao, "Predictive modeling of undeformed chip thickness in ceramic grinding", *Int. J. Mach. Tool. Manuf.* 56, 59–68, 2012.

[20] M. Nomura, *Study of ultrasonic vibration assisted internal grinding of small holes*, Doctoral Dissertation, Tohoku University, 2005. (In Japanese)

[21] M. Fujimoto, Y. Wu, J. Cao, "High Precision Ultrasonically Assisted Internal Grinding (UAIG) of Difficult-to-machining Materials using Metal Bonded Diamond Wheels," The 6th International Conference on Leading Edge Manufacturing in 21st Century, 2011.

[22] M. Nomura, Y. Wu, M. Kato, T. Kuriyagawa, "Effects of ultrasonic vibration in truing and dressing of CBN grinding wheel used for internal grinding of small holes", *Key Eng. Mater.* 291-292, 183-188, 2005.

[23] Y. Wu, M. Nomura, Z. Feng and M. Kato, "Modeling of Grinding Force in Constant-Depth-of-Cut Ultrasonically Assisted Grinding", *Mater. Sci. Forum* 471-2, 6-10, 2004.

[24] Ioan, D. Marinescu, Mike Hitchiner, Eckart Uhlmann, W. Brian Rowe, Ichiro Inasaki, *Handbook of Machining with Grinding Wheels*, CRC Press, Taylor and Francis Group, 6000 Broken Sound Parkway NW, 14, 2007.

[25] S. Malkin, T. W. Hwang, "Grinding Mechanisms for Ceramics", *CIRP Annals - Manufacturing Technology*, 45, 2, 569-580, 1996.

[26] B. Denkena, T. Friemuth, M. Reichstein, "Potentials of different process kinematics in micro grinding", *Ann. CIRP* 52, 463-466, 2004.

[27] H. Mult, G. Spur, S. Holl, "Ultrasonic assisted grinding of ceramics", *J. Mater. Process. Technol.* 62, 287-293, 1996.

[28] K. Suzuki, T. Makizaki, S. Miake, T. Uematsu, "Study on ultrasonic elliptic vibration grinding: 1st report- proposal of an ultrasonic elliptic vibration grinding wheel by utilizing a stator of an ultrasonic motor", *J. Jpn. Soc. Abras. Technol.* 42, 448-653, 1998. (in Japanese)

[29] K. Suzuki, T. Uematsu, T. Makizaki, M. Iwai, "Ultrasonic grinding utilizing a stator of an ultrasonic motor", *Proc. of JSME Conference Manufacturing and Machine Tool,* Mishima, Japan, 183-184, 2000.

[30] Y. Yan, B. Zhao, J. Liu, "Ultraprecision surface finishing of nano-ZrO_2 ceramics using two-dimensional ultrasonic assisted grinding", *Int. J. Adv. Manuf. Technol.* 43, 462-467, 2009.

[31] Z. Liang, Y. Wu, X. Wang, W. Zhao, "Two-dimensional ultrasonically assisted grinding technique of monocrystal silicon", *Proc. of the 5th International Conference on Leading Edge Manufacturing in 21st Century,* Osaka, Japan, 535-540, 2009.

[32] Y. Wu, Z. Liang, W. Lin, "Surface Formation Characteristics in Elliptical Ultrasonic Assisted Grinding of Monocrystal Silicon", *Inter. J. Nanomanuf.* 7, 189-198, 2011.

[33] Z. Liang, X. Wang, Y. Wu, W. Zhao, "Elliptical Ultrasonic Assisted Grinding (EUAG) of Monocrystal Sapphire- Surface Formation Characteristics", *Adv. Mater. Res.* 126-128, 367-372, 2010.

[34] Z. Liang, Y. Wu, X. Wang, W. Zhao, "Development of a two-dimensional ultrasonic vibration assisted grinding technique of monocrystal silicon", *Chn. J. Mech. Eng.* 13, 192-199, 2010. (in Chinese)

[35] Y. Wu, Z. Liang, X. Wang, W. Lin, "Elliptical Ultrasonic Assisted Grinding (EUAG) of Monocrystal Sapphire- Wear Behaviors of Resin Bond Diamond Wheel", *Adv. Mater. Res.* 126-128, 573-578, 2010.

[36] V. V. Rogov, Y. Filatov, W. Kottler and V. P. Sobol, "New technology of precision polishing of glass optics", *Opt. Eng.* 40, 1641–1645, 2001.

[37] Y. D. Filatov, O. Y. Filatov, G. Monteil, U. Heisel and M. Storchak, "Bound-abrasive grinding and polishing of surfaces of optical materials", *Opt. Eng.* 50, 063401, 2011.

[38] B. E. Gillman and S. D. Jacobs, "Bound-abrasive polishers for optical glass", *Appl. Opt.* 37, 3498–3505, 1998.

[39] W. Xu, X. Lu, G. Pan, Y. Lei and J. Luo, "Ultrasonic flexural vibration assisted chemical mechanical polishing for sapphire substrate", *Appl. Suf. Sci.* 256, 3936-3940, 2010.

[40] P.-L. Tso and C. Tung, "Study on ultrasonic-assisted chemical mechanical polishing", *Int. J. Abra. Technol.* 4, 132-139, 2011.

In: Ultrasonics: Theory, Techniques and Practical Applications ISBN: 978-1-62257-685-2
Editors: Hanako Ayabito and Mitsuko Katsukawa

Chapter 5

DYNAMICS AND KINETICS OF SUPERCOOLED LIQUIDS UNDER ULTRASONIC FIELD

Hiroshi Abe
Department of Materials Science and Engineering,
National Defense Academy, Japan

ABSTRACT

In simple system (metallic sodium, benzene and naphthalene), supercooled states were investigated by ultrasonic measurements. Frequency dependent shear loss modulus can detect relaxation process sensitively in the supercooled liquid of metallic sodium. The dynamic heterogeneity might correspond to a dynamic cluster at the supercooled state. In contrast to the monatomic system, benzene and naphthalene possess anisotropic interactions intrinsically due to anisotropic molecular shapes.

The peculiar feature is characterized by a competition between π-π repulsive interaction and molecular packing efficiency. On balancing in the local circumstance, supercooled benzene is coupled with the ultrasonic fields. Furthermore, the ultrasonic irradiation effect is enhanced by thermal treatments such as quenching. Kinetics during a non-equilibrium state can extract hidden information about the nucleation process, where dynamic heterogeneous fluctuations in the supercooled liquids are dominant in dissipative system.

Keywords: Shear impedance; Ultrasonic irradiation effect; Supercooled liquid; Free-energy landscape; Dynamic heterogeneity; Dissipative system; Dynamic embryo; Frozen nucleus

1. INTRODUCTION

In the 21st century, 'complexity' is in great progress in order to examine a relationship between complicated structure and multifunctional property. Basically, 'complexity' in physics is divided into two: one is an element in random media described by many degrees of freedom, and the other is multicomponent in system. For instance, the former is supercooled liquid or glass in simple system. The latter corresponds to protein in solution as soft matter. In particular, the protein structure and its hyper structure are strongly related with intrinsic functional properties. Adjustment (feedback mechanism), flexibility (geometrical effect) and selectivity (switching mechanism), etc. in the protein system is developed to the neural network, whose functions are amount of information, information memory, information selectivity and information transfer. In this article, we focus on supercooled liquids in simple atomic and molecular systems to clarify the fundamental dynamic and thermal property using ultrasonic technique. As the first step, order and disorder states in condensed matter are classified by three elements, that is, (i) chemical, (ii) bond (the constant coordination number) and (iii) topological factors (the variable coordination number) as shown in Figure 1.

In anisotropic molecular or spin system, orientational order/disorder is additionally considered. On the basis of the above disorder factors, glass transition has been discussed for a long time. A glass transition relating to the supercooled liquid is reviewed in the literature [1]. Dynamic properties such as experimentally obtained α- and β- relaxation processes are still a big problem. In addition, the supercooled liquid states are classified into 'strong' and 'fragile' ones [2].

Two kinds of features are well distinguished on viscosity plot ('Angell plot'), where temperature is normalized by glass transition temperature (T_g). Theoretically, the idea of a free-energy landscape was provided to describe the difference between 'strong' and 'fragile' supercooled liquids. Furthermore, molecular dynamics (MD) simulations were carried out by introducing two order parameters [3].

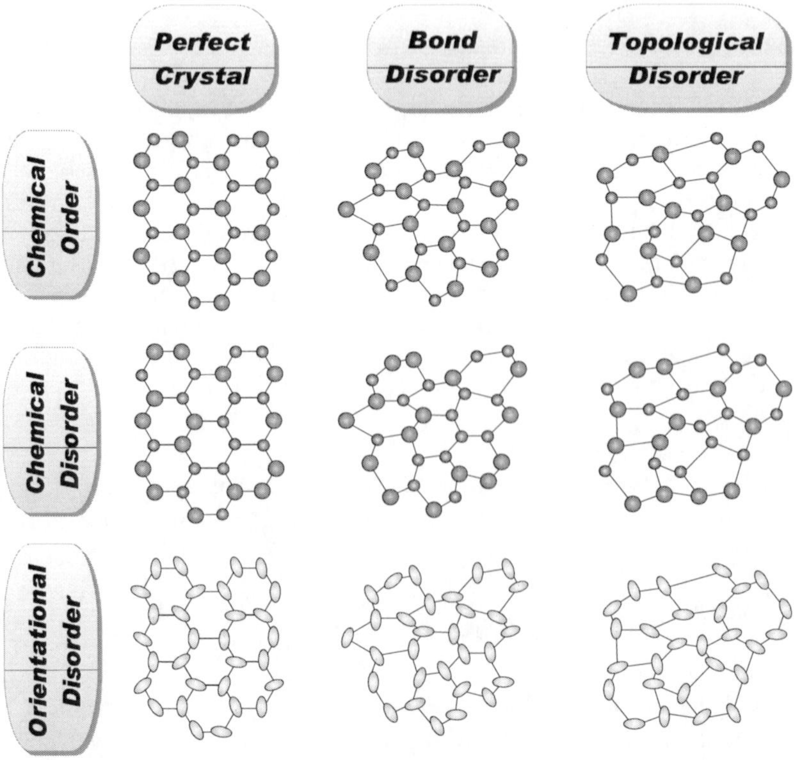

Figure 1. Schematic order and disorder states in condensed matter.

Two-dimensional (2D) glass-forming model having strong anisotropic interactions tells how important geometrical frustrations are in the complex system. For instance, local five-fold symmetry appeared partially in the supercooled liquid phase, where the pentagon avoids crystal nucleation. Macroscopically, the dynamic heterogeneity is discussed with free-energy landscape. Further 3D MD analysis in supercooled monatomic liquid showed interesting behaviors [4].

Even below T_g, dynamic heterogeneity is explained by liquid-like atoms inside glassy matrices. In liquid of metallic sodium, MD simulation was carried out considering mode-coupling theory [5]. In the simulation box, 686 Na atoms were quenched from 423 K down to 23 K. The calculated T_g is much closer to the real T_g that was determined by the calorimetric measurement. Furthermore, the neutral hard sphere-model using mode-coupling theory was extended into the charged hard-sphere one [6].

The model is described by the short-range effect such as geometrical packing and the long-range effect of Columbic charge. On the other hand, nucleation process was demonstrated by the MD simulations [7]. A various kinds of Na clusters appeared in the in a system consisting of 10,000 Na atoms. In the supercooled liquid, the transition process from the icosahedral cluster to the deformed bcc nanostructure was virtually realized in the simulations. Finally, the perfect bcc structure as a solid phase is formed with vanishing the region of the supercooled liquid state.

Benzene (C_6H_6) is simple organic molecule having π electron. Nonpolarity in benzene is derived from its perfectly symmetric molecular structure. The liquid [8, 9] and crystal structures [10, 11] of benzene suggest that π-π repulsive interactions are not ignored both in liquid and crystal. Also, the crystal structure of naphthalene ($C_{10}H_8$) [12, 13] satisfies the above tendency. In the same manner with benzene, naphthalene has no polarity at all. Apart from liquid or crystal, molecular glasses were investigated in benzene, naphthalene and anthracene. In simple organic system, the deposition method on cold-metal substrates was effective to prevent the crystallization. The thin films of benzene [14], naphthalene [15] and anthracene [16] were amorphized by rapidly quenching. For instance, benzene glass was obtained below 30 K (temperature of cold-metal substrate). On heating, the benzene glass crystallized at 60 K. In addition, the crystallization temperatures of naphthalene and anthracene films were found to be 105 K and 200 K, respectively. In this article, the supercooled liquid states in simple atomic and molecular systems are highlighted. Dynamic property of supercooled sodium implies the existence of the specific cluster. In molecular supercooled liquids, anomalous nucleation and growth process is easily coupled both with external ultrasonic field and thermal treatments.

2. EXPERIMENTAL

2.1. Shear Impedance Method

In order to measure the shear impedance, ac-cut quartz crystal was used. The pulse modulator and receiver is Model 7700 (Matec Co.), whose frequency (ν) covers the range from 1 to 20 MHz. R.F. Plug-IN (Model 755, Matec Co.) is inserted into the Model 7700. G^*

(= G' + iG'') is the complex shear modulus and C_s^* (= C_s' + iC_s'') is the complex shear velocity. Furthermore, the relationship between G^* and $C_s'$is given by [17-20],

$$\rho C_s'^2 = \frac{G'}{2}\left\{1 + \sqrt{1 + (G''/G')^2}\right\}, \tag{1}$$

where ρ is the density of a specimen. $\rho C_s'^2$ is also obtained from the experimental results as follows [21, 22],

$$\rho C_s'^2 = \frac{z_q^2}{\rho}\tanh\left\{\frac{D_\omega(T)}{40\log_{10}e}\right\} \tag{2}$$

where z_q is the impedance of the quartz delay line ($z_q = 8.78\times10^5$ g/cm$^2\cdot$ sec.), $D_\omega(T)$ is the difference between the dB loss/echo for liquid-quartz interface and that for air-quartz interface. Metallic sodium (Rare Metallic Co), whose purity is 99.9 %, was used in this study. To prevent metallic sodium from the oxidation, we use a set of the double containers [23]. Smaller one is the aluminum container, which is filled with liquid paraffin oil. Metallic sodium is chemically stable both against aluminum and liquid paraffin. We fixed this smaller container inside the larger container, which is made of a heat insulating material and is filled with the silicone oil. Temperature was controlled within ±0.5 °C using the above double containers and the precise temperature controller.

2.2. Liquid-Immersion Through-Transmission Ultrasonic Technique

High purity benzene for spectroscopy (Wako Pure Chemical Co.) was selected in this experiment. Liquid benzene is put into the sample cell, whose diameter is 40 mm and thickness is 10 mm. To reduce ultrasonic absorption and reflection on the windows of the sample cell, both transmission sides were shielded by poly vinylidene chloride films, whose thickness is 20 μm. Temperature of the sample was monitored by inserting a thermocouple into the sample cell. By the conventional ultrasonic pulse transmission method, the ultrasonic longitudinal absorption and longitudinal wave velocity were measured. Sine waves are generated by a pulse function generator (8116A, Hewlett Packard Co.) Two piezoelectric transducers (1 MHz PZT, Ultrasonic Engineering Co.) were used for generating and receiving longitudinal waves. The voltage applied to PZT is 10 V. The data of ultrasonic waves were collected using a digital storage oscilloscope (VC-6065, Hitachi Co.) The quenching method is that the liquid sample at the room temperature, T_{rt} (= 20 °C), is put directly into the ethanol bath (Heto, Type CB7) [24]. The quenched temperature (T_Q) is set to the range from 0 down to −20 °C. Temperature stability inside the ethanol bath is within 0.2 °C. Ultrasonic longitudinal absorption is given by [25, 26],

$$L = 20\log_{10}\frac{\xi_0}{\xi} + 20\log_{10}\frac{4zz_0}{(z+z_0)^2}, \tag{3}$$

where ξ_0 and ξ are amplitude of the ultrasonic wave without the sample (ethanol only) and with the sample, respectively. The second term of Eq. (3) reveals the correction of twice reflections on the organic films. z (=ρC_L) and z_0 (=$\rho_0 C_0$) are acoustic impedance of the sample and ethanol, respectively. C_0 is the velocity of liquid ethanol and C_L is the velocity of the sample. ρ and ρ_0 are the densities of benzene and ethanol.

2.3. Ultrasonic Burst Pulse Irradiation

Naphthalene (99.0% Kanto Chemical Co., Inc.) was purified by the zone melting method. By 100 solvent cycles, the sample was purified up to 99.8% [27]. The sample cell used for ultrasonic measurement was a brass cylinder, whose thickness was 5 mm and diameter was 50 mm. Glass plates, whose thickness was 0.4 mm, were attached on the brass cylinder mechanically to minimize ultrasonic absorption.

A thermocouple (chromel-alumel) was set inside the sample cell as a temperature monitor. Quenching and slow cooling experiments were performed using water bath (SR-31, Yamato Co.), where temperature was controlled within 0.1 K.

The burst sine waves of ultrasonic sound were generated by a pulse function generator (8116A, Hewlett Packard Co.) Acoustic pressure is proportional to applied power, P, in an ultrasonic transducer, which is controlled by the bust values. Two sets of 1 MHz PZT were used for generating and receiving longitudinal waves. The data collection for ultrasonic waves was performed using a digital storage oscilloscope (VC-6065, Hitachi Co.) In order to investigate whether or not the amorphous state exists, we carried out X-ray diffraction experiments using the samples obtained by ultrasonic irradiation. Mo radiation was used as the primary beam. Laue photographs using imaging plates (IPR-420, Mac Science Co.) were taken using a 2 kW X-ray generator (Rigaku Co.).

3. Experimental Results and Discussions

3.1. Relaxation Process in Supercooled Liquid of Sodium

Shear impedance method has been established for molecular and dilute polymer solution, and so on. It is worth to check its validity to simple monatomic system, excluding chemical order/disorder. Also, orientational order/disorder as seen in anisotropic shaped molecules is negligible in the monatomic system.

Figure 2 shows the temperature dependence of $\rho C_s'^2$ at 10.0 MHz (=ν). Blue and red closed circles reveal the cooling and heating process, respectively. The cooling and heating rates were 0.5 °C/min. On cooling, solidification occurred at around 50 °C. Here, we define solidification temperature as T_s (= 50 °C). Since melting point (T_m) of sodium is 97.9 °C, supercooled liquid of sodium existed below T_m. Simultaneously, it was found that obvious hysteresis appeared in $\rho C_s'^2$ at the thermal cycle. Degree of supercooling is defined by introducing reduced temperature, Θ. Θ is expressed by,

$$\Theta = \frac{T_m - T_s}{T_m}, \tag{4}$$

where temperature unit is K. In this experiment, Θ_{Na} was found to be 0.129. Temperature region of the supercooled liquid in sodium is relatively large in spite of simple monatomic system. Generally, it is interpreted that creation of polyhedral clusters in the supercooled liquid prevents crystal nucleation. In case of metallic sodium, theoretically predicted icosahedron clusters [7] might exist partially. Crystal nucleation is geometrically suppressed from the specific cluster, even though temperature is deeply far from T_m.

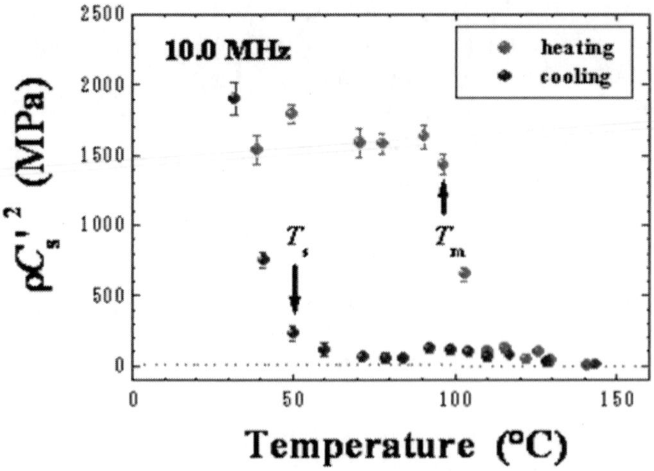

Figure 2. Temperature dependence of $\rho C_s'^2$ at 10.0 MHz. Blue and red closed circles represent cooling and heating process, respectively. T_m is the melting point of metallic sodium.

Next, we focus on the frequency (ν) dependence of $\rho C_s'^2$ having information about relaxation time. ν dependence of $\rho C_s'^2$ consists of G' and G'' in Eq. (1). Analytically, we cannot obtain frequency dependence of G' and G''. However, G' and G'' are numerically decomposed using the Kramers-Kronig relation and Monte Carlo (MC) method. In addition, we assume that $G''(\nu)$ is represented approximately by a polynomial, which is provided by $p_0 + p_1 f + p_2 f^2 + p_3 f^3$. $G'(\nu)$ is connected with the $G''(\nu)$, which is satisfying with the Kramers-Kronig relation. The convergence condition for the MC calculations is a reliability factor (R), which is provided by,

$$R = \frac{\Sigma\left|(\rho C_s'^2)^{cal} - (\rho C_s'^2)^{obs}\right|}{\Sigma(\rho C_s'^2)^{obs}}. \tag{5}$$

The reliability factor converged to almost constant value at around 10^7 MC steps, where the R is less than 2 %. Using the calculated p_0, p_1, p_2 and p_3 parameters on cooling, the decomposed $G''(\nu, T)$ is revealed in Figure 3. Above T_m, the G'' is very small, while the G''

increases rapidly below T_s. As an intermediated state between them, dynamic property at the supercooled state is entirely different both from normal liquid and solid states.

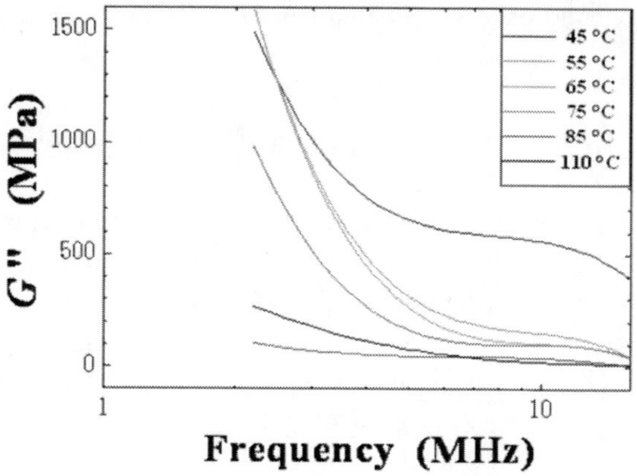

Figure 3. Temperature and frequency dependence of *G*” on cooling in metallic sodium.

In principle, a dynamic property is analyzed using a generalized Maxell model, which contains a relaxation time, τ. That is given by [28],

$$G''(\omega) = \int_0^{\infty} H(\tau) \frac{\omega\tau}{1+\omega^2\tau^2} \frac{d\tau}{\tau}. \tag{6}$$

$H(\tau)$ is the relaxation spectrum. Therefore, drastic change of *G*” spectra at $T_s < T < T_m$ corresponds to transitional relaxation process of the dynamic clusters. Here, we summarize the features of $\rho C_s'^2$ and *G*” (Table 1) as a probe because their responses are so different at each states. The former $\rho C_s'^2$ is a typical macroscopic shear response with traveling in liquid-like media. Hence, C_s' is a proper probe to determine whether a sample is liquid-like or solid-like state. While, the relaxation process in the supercooled liquid is relating to dynamic local fluctuations, which depends on size, population and lifetime of the dynamic polyhedral clusters. The anomalies in *G*” spectra are seen not only in supercooled liquid of monatomic system but also in dilute polymer solutions (hierarchical structure) [29], liquid crystal (self-organized structure), gels (network structure) and protein solutions (hyperstructure), and so on. Furthermore, dynamical shear-modulus measurements were applied into glass-forming molecular liquids [30].

Table 1. Sensitivity at each state both of $\rho C_s'^2$ and *G*’. Removing an exceptional case, sensitivity as a probe is expressed roughly as 仢 > 仜 > 仮 > 亊

	Solid	Supercooled liquid	Liquid
$\rho C_s'^2$ transverse waves	◎	Δ	×
G” mechanical loss	○	◎	×

In the frequency-resolved mechanical spectra, α- and β-relaxation processes in the supercooled liquids were examined associating with glass transition. The peak of shear loss modulus shifted drastically just above T_g. We stress, here, that the shear impedance method is a significant tool both for macroscopic elastic property and local relaxation process.

3.2. Ultrasonic Irradiation and Quenching Effect in Supercooled Benzene

External longitudinal ultrasonic field causes compression and expansion alternatively in a liquid sample. In pressure variable circumstance, liquid density is easily influenced by molecular orientational order/disorder of anisotropic shaped molecules. Since benzene has no polarity, dipole-dipole interaction is ignored. Dynamic heterogeneity such as high and low density enhances the intrinsic property at the supercooled state. Temperature dependence of longitudinal absorption on slow cooling, $L_{SC}(T)$, is shown in Figure 4(a).

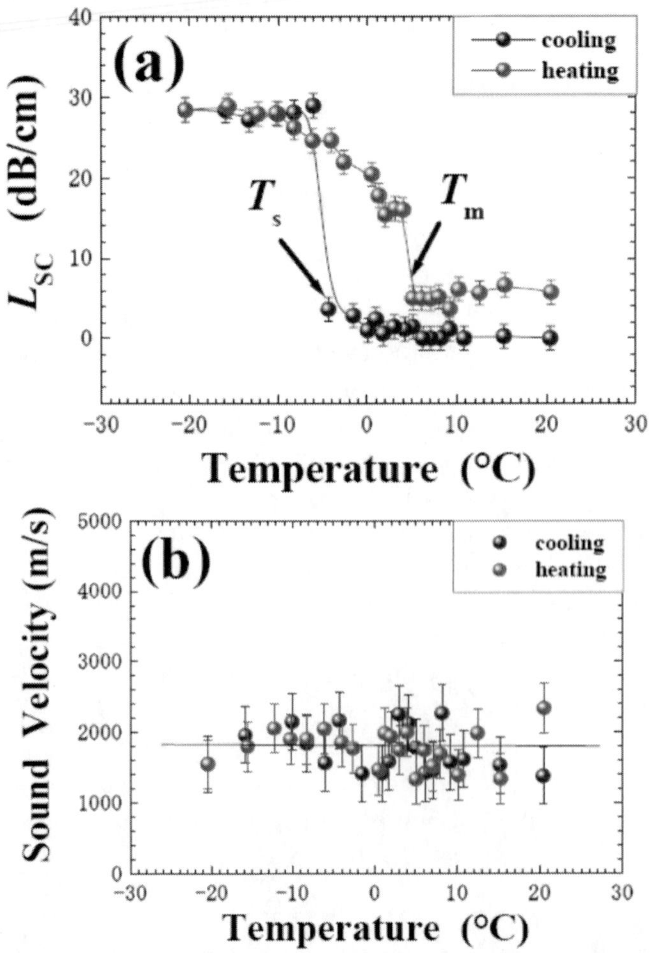

Figure 4. Temperature dependence of (a) longitudinal absorption (L_{SC}) and (b) longitudinal wave velocity in benzene. Blue and red closed circles are corresponding to the points for slow cooling and heating process, respectively.

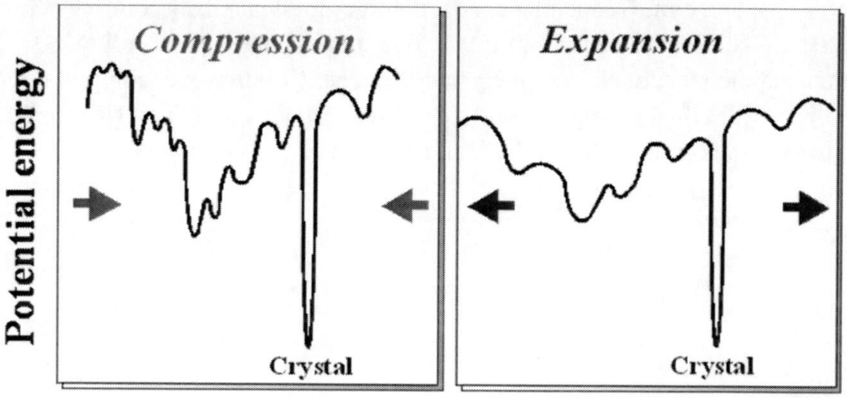

Figure 5. Pressure dependent free-energy landscape. Under longitudinal ultrasonic field, free-energy is modified on space and time.

Here, the cooling rate is 0.08 (°C/min) and the heating one is 0.12 (°C/min). Abrupt increasing/decreasing of the longitudinal absorption is driven by the state changes. On the slow cooling, the solidification of benzene occurred at -5.0 °C (T_s), which is below T_m (= 5.5 °C). The supercooled benzene appeared at $T_s < T < T_m$. Θ of benzene (Θ_{C6H6}) is calculated to be 0.038. Compared with Θ_{Na}, degree of supercooling of benzene is relatively small. To compare Θ_{C6H6} with Θ_{Na}, we must distinguish influence of longitudinal wave from that of transverse wave (shear impedance method). By pressure variant effect under the longitudinal wave, free-energy should be modified by alternative longitudinal compression/expansion. We speculate the free-energy landscape under the longitudinal ultrasonic field (Figure 5).

The schematically illustrated potential energy is based on high pressure researches. Molecular mobility or confinement under low and high pressure is represented by energy barrier between local minima.

In partially expanded region, pressure becomes relatively lower. Then, loosely packed molecules are translating and rotating easily. While, in dense packed region due to high pressure, the molecular mobility is restricted. Particularly in geometrically anisotropic molecules, orientational order is promoted to reduce volume effectively. Since the well-arranged region is similar to a periodic molecular array in crystal, events of crystal nucleation might increase. The compression/expansion regime can explain small Θ_{C6H6} under the longitudinal ultrasonic field. Moreover, another significant finding is that precursor phenomenon of solidification is seen in the $L_{SC}(T)$. Close to T_s, gradual increment of $L_{SC}(T)$ was detected within experimental errors. $L_{SC}(T)$ is sufficiently sensitive for density fluctuations such as a pre-transition stage. In contrast to hysteresis of the longitudinal absorption, longitudinal wave velocity was almost constant at the thermal cycle. (Figure 4(b)).

As another approach to study relaxation process in the supercooled liquid, a quenching experiment is indispensable. In time resolved experiments, hidden information about dynamic fluctuations is extracted. In addition, ultrasonic irradiation at the non-equilibrium state is a useful technique to detect local diffusion, heterogeneous fluctuations and other collective motions on the random media. Kinetic experiments using quenched benzene were carried out at the fixed temperature. Time was set to be 0 when the sample temperature achieved to the holding temperature (T_Q).

Figure 6 displays time dependence of the longitudinal absorption in the quenched benzene at 0 °C. The curious points in quenching are summarized as follows; (i) ultrasonic waves becomes quite unstable in the quenched benzene, (ii) after quenching, the longitudinal absorption in the quenched benzene, $L_Q(T_Q, t)$, is larger than $L_{SC}(T)$ and (iii) the longitudinal absorption has the maximum value at the incubation time, t_{inc}.

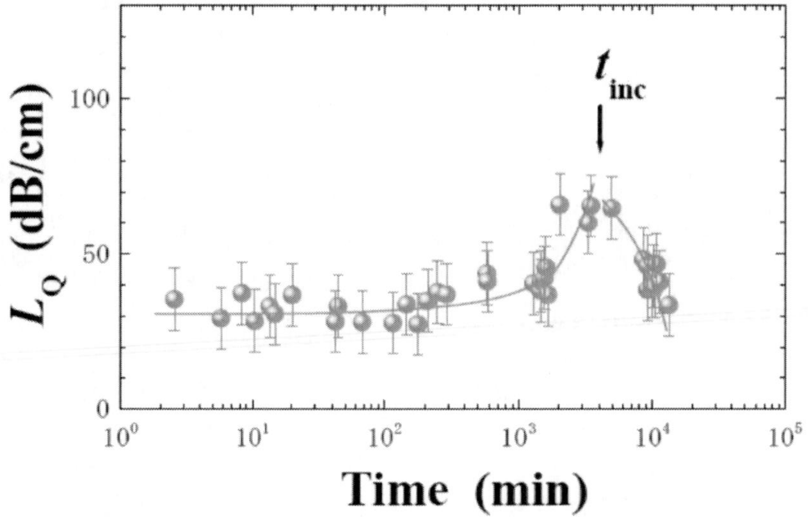

Figure 6. Time development of longitudinal absorption in quenched benzene, L_Q, at 0 °C. L_Q has the maximum value at the peculiar time, t_{inc}.

Three anomalies are relating to ultrasonic assisted nucleation as discussed previously about $L_{SC}(T)$. The quenching feature (i) suggests randomly appeared or disappeared dynamic 'embryo' [31]. The concept of the 'embryo' is that molecular aggregation of the 'embryo' is similar to periodic array of crystal (Figure 7).

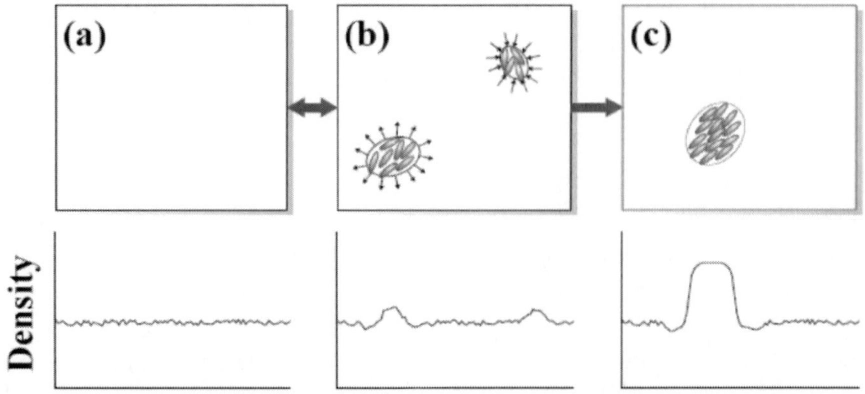

Figure 7. Schematic molecular aggregations and density fluctuations at (a) homogeneous liquid, (b) reversible 'dynamic embryo' in supercooled liquid and (c) irreversible 'frozen nucleus' inside the supercooled liquid. For simplicity, molecules at average density are not drawn.

Creation and annihilation of the 'embryo' randomly occurs. Once the dynamic 'embryo' exceeds over a critical size, it changes to static 'frozen nucleus', whose structure is exactly the same as crystal one. The 'frozen nucleus' never transforms back to the dynamic 'embryo'. By quenching, a lot of the dynamic 'embryos' appeared in the quenched benzene.

The phenomenon (i) is equivalent to acoustic speckle generated by ultrasonic waves traveling through random media [32]. The acoustic speckle is based on the idea that some propagating wave interferes with others through multiple scattering paths. Dynamics of heterogeneous materials was evaluated by theoretical approach using diffusing acoustic wave spectroscopy.

A large number of dynamic 'embryos' also explain the larger acoustic attenuation ($L_Q(T_Q, t) > L_{SC}(T)$) described in the feature (ii). Ultrasonic waves could be scattered by the randomly located and created 'embryos'. Hence, additional acoustic attenuations imply the excess 'embryos' in the quenched benzene. Focusing on the feature (iii), we apply the idea of the 'embryo' into the maximum value of $L_Q(T_Q, t_{inc})$.

By a combination of quenching and externally injected vibrational energy, the pressure assisted dynamic 'embryos' are excited extensively. But the randomly distributed 'embryos' cannot compensate the geometrical frustration. The frustration affects as pinning effect in the quenched benzene. Then, the dynamic 'embryos' might be accumulated with a lapse of time. Close to t_{inc}, population of the 'embryo' becomes larger. At around t_{inc}, a fully developed 'embryo' starts to transform to a 'frozen nucleus'. At the late stage ($t_{inc} < t$), decreasing 'embryo' and growth/coarsening of crystal domain make $L_Q(T_Q, t)$ smaller. A fully-grown crystal domain has small specific surface area of the domain, since the specific surface area is inverse proportional to its volume. Ultrasonic waves are less scattered at the late stage. Consequently, the peak of $L_Q(T_Q, t)$ is formed by a balance between 'embryo' strongly coupled with ultrasonic waves ($t < t_{inc}$) and domain growth/coarsening promoted by ultrasonic irradiation ($t_{inc} < t$).

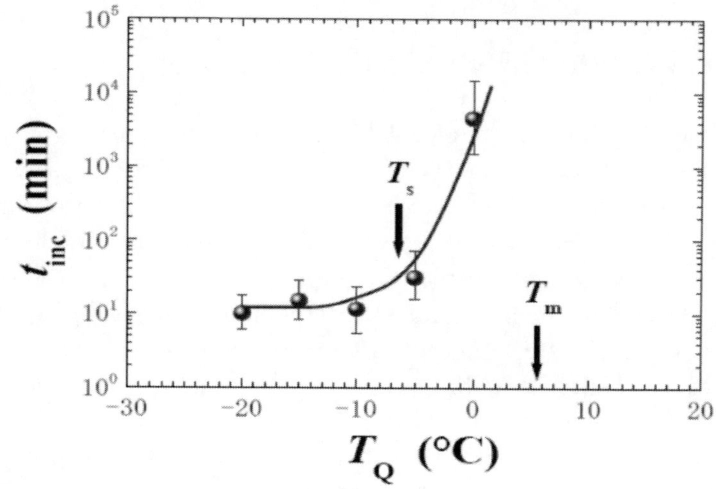

Figure 8. Incubation time, t_{inc}, as a function of quenching temperature in quenched benzene. The waiting time is determined by the maximum values of longitudinal absorption (Figure 6). t_{inc} diverges close to T_m. The solid line shows the calculated incubation time regarded as the thermal activated process.

Kinetics during incubation time is influenced by quenching depth, ΔT_Q (= T_{rt} - T_Q), which stands for degree of a non-equilibrium state. For example, temperature [33] and pressure [34] jump experiments were effective to clarify the nucleation and growth process. In particular, incubation time was connected with nucleation barrier thermodynamically.

In this study, $L_Q(\Delta T_Q, t_{inc})$ was measured systematically to analyze incubation time. Figure 8 reveals ΔT_Q dependence of t_{inc}. It is figured out that longer t_{inc} was observed with increasing the holding temperature. In the classic nucleation model [35], thermodynamic scheme consists of the following three factors: (i) activation energy, (ii) nucleation probability and (iii) ergodic theorem. A nucleation barrier, $\Delta E(T)$, as thermally activated process is given approximately by [36, 37],

$$\Delta E(T) = \alpha + \beta T + \gamma T^2. \tag{7}$$

In addition, the nucleation probability, p, is calculated using the Boltzmann factor. Thus, t_{inc} is calculated taking into account ergodic theorem and is provided by,

$$\tau_{inc} \propto 1/p \propto \exp\{\Delta E(T)/k_B T\} \tag{8}$$

where k_B is Boltzmann constant. α, β and γ in Eq. (7) are fitted by the least square method. The solid line in Figure 8 shows the calculated t_{inc}. Theoretically defined incubation time is the waiting time until a 'frozen nucleus' occurs. It should be noticed, in the supercooled liquid, that $\Delta E(T)$ is time- and space-average of the nucleation barrier on a free-energy landscape. The classic nucleation model is formalized based on the idea of the mean field approximation (homogeneous media). The interaction between the dynamic 'embryos' is not taken into account.

3.3. Ultrasonic Power Effect in Supercooled Naphthalene

Naphthalene is more anisotropic molecule than benzene. The planar molecule has no polarity. Due to the geometrical effect, T_m of naphthalene is 80 °C. On slow cooling and heating, $L_{SC}(T)$ as a function of temperature is shown in Figure 9. The cooling and heating rate is 0.4 (°C/min). Under longitudinal ultrasonic field, naphthalene solidified at 76.3 °C (=T_s). The thermal hysteresis of naphthalene was quite small (Θ=0.009). A significant finding in $L_{SC}(T)$ of naphthalene is that the longitudinal absorption upon heating gradually increased toward T_m. Surprisingly, the $L_{SC}(T)$ value is larger than twice value at 50 °C (solid phase). This tendency is not seen both in metallic sodium (Figure 2) and benzene (Figure 4(a)) upon slow heating. One of possible ideas is that morphological change of crystal domain is enhanced in melting. In deformation of crystal domains, dynamic heterogeneous fluctuations might develop close to T_m. Since the melting behavior in naphthalene is something strange, we checked T_s with or without the ultrasonic irradiation. Figure 10 exhibits ultrasonic power dependence of T_s in different cooling rate (2 °C/min). Here, T_s was determined with ultrasonic absorption (Θ=0.022) though T_s under no ultrasonic irradiation (Θ=0.028) was determined by the time dependence of the monitored temperature. Open circle in Figure 10 reveals T_s under no ultrasonic irradiation. Obviously, T_s under the ultrasonic field became higher than T_s without the ultrasonic irradiation.

The ultrasonic field assists to crystallize on slow cooling. However, T_s has no ultrasonic power (P) dependence. At $0.03 < P < 0.30$, T_s was almost constant within experimental errors. No power relation of T_s still remains unclear. Next is kinetics in the supercooled naphthalene with the isothermal holding ($T > T_s$).

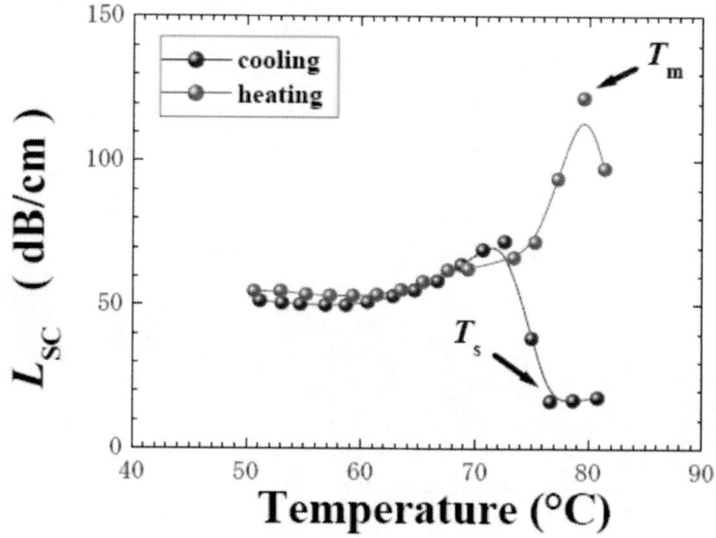

Figure 9. Longitudinal absorption as a function of temperature in naphthalene. Blue and red closed circles reveal slow cooling and heating process, respectively.

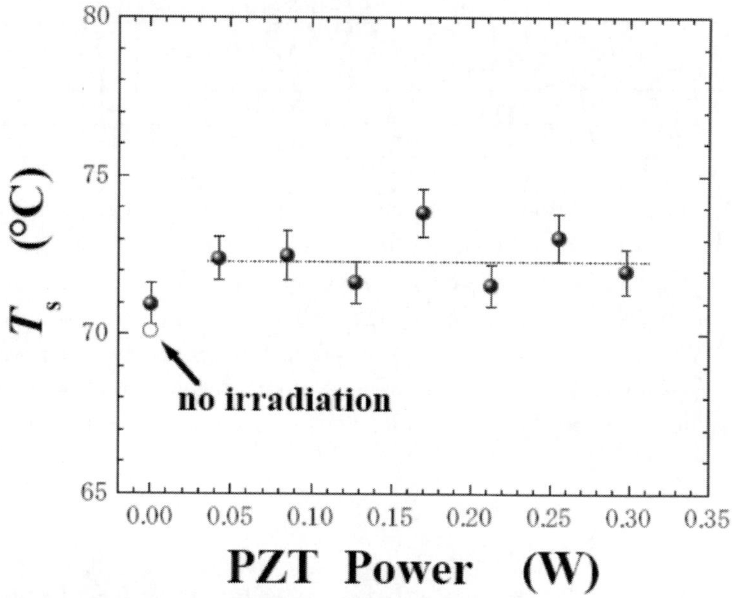

Figure 10. Solidification temperature (T_s) as a function of irradiated ultrasonic power. Under no ultrasonic irradiation, T_s decreases as denoted by the open circle.

Figure 11(a) is one of examples under isothermal holding combined with ultrasonic irradiation. The ultrasonic irradiation power is 0.064 W. Thermal trace was plotted in Figure 11(b). Certainly, temperate was held isothermally at 77.1 °C. The ultrasonic waves were so stable over the whole time region. The incubation time (t_{inc}) was observed at fixed temperature in spite of small hysteresis of the thermal cycle. After t_{inc} of 53.3 min, $L_Q(T_Q, t)$ started to increase and the time dependence of $L_Q(T_Q, t)$ obeys the grow model [38]. The solid line is calculated by the modified Kolmogorov equation. By introducing the screening effect to avoid overlapping of each domain, the model is more realistic in growth process. If we assume that $L_Q(T_Q, t)$ is proportional to the number of domains without coarsening, the supercooled naphthalene is regarded as a prototype of the nucleation and growth process. Besides, the normal behavior in the quenched naphthalene emphasizes the outstanding behaviors of $L_Q(T_Q, t)$ in the quenched benzene. Including the slow heating results, the anomalies of benzene and naphthalene are listed for a comparison (Table 2). Here, η is the anisotropic geometrical factor to express the molecular shapes quantitatively. When molecular size is provided by 3D lengths ($a < b < c$), η is obtained simply as follows,

$$\eta = \frac{b}{a}\frac{c}{a}. \tag{9}$$

Comparing the experimental results with each factor, we confirm that concept of dynamic 'embryo' can explain the anomalies in the quenched benzene.

Table 2. Comparison between benzene and naphthalene. η and Θ are anisotropic shape factor of molecules and reduced hysteresis, respectively

			athermal	isothermal		
	η	Θ	L_{SC} (heating)	*ultrasonic* wave	L_Q $t < t_{inc}$	L_Q $t_{inc} < t$
Benzene	13.9	0.038	constant	unstable	increase	decrease
Naphthalene	18.9	0.009	increase	stable	constant	constant

Now, we have a big question: is there any irradiation power dependence of t_{inc} in the quenched naphthalene? On slow cooling, no irradiation power dependence of T_s was observed. Thus, we carried out the systematic experiments under both ultrasonic irradiation and isothermal holding, as shown in Figure 12. Both ultrasonic irradiation and isothermal holding above T_s yielded two interesting results. One is that t_{inc} diverged with increasing ultrasonic power. The other is that t_{inc} increased with increasing holding temperature. To interpret the complicated experimental results, we analyze the t_{inc} using Eqs. (7) and (8). Here, Eq. (7) is modified as,

$$\Delta E(T, P) = \alpha + \beta T + \gamma P, \tag{10}$$

since t_{inc} is proportional to P. However, the above phonological approach is not sufficient to explain our experimental results. Therefore, a coupling effect between local pressure and holding temperature is necessary to coincide with the experimental results. We introduce the coupling constant, δ with replacing the last term in Eq. (10). The improved $\Delta E(T)$ is

$$\Delta E(T, P) = \alpha + \beta\, T + \delta\, TP. \tag{11}$$

The solid lines in Figure 12 show the incubation time calculated by our simple model. In case of quenched naphthalene, ultrasonic irradiation cannot promote nucleation process. For nucleation process, t_{inc} become longer proportionally to local pressure.

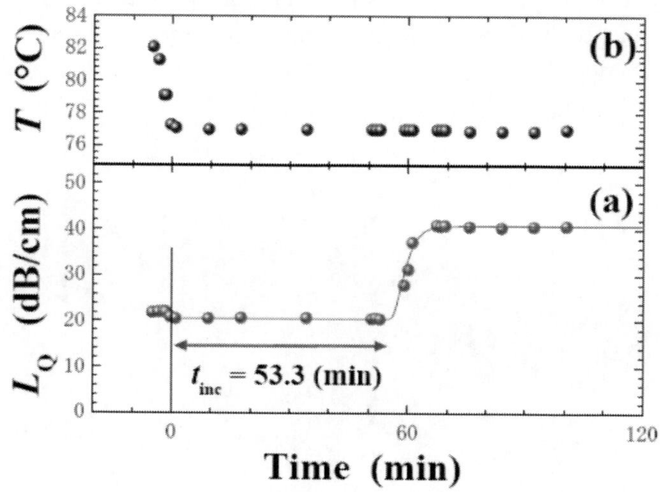

Figure 11. Time dependences of (a) longitudinal absorption and (b) temperature trace in the quenched naphthalene. Red curve in Figure 11(a) is calculated by the improved growth model (see text).

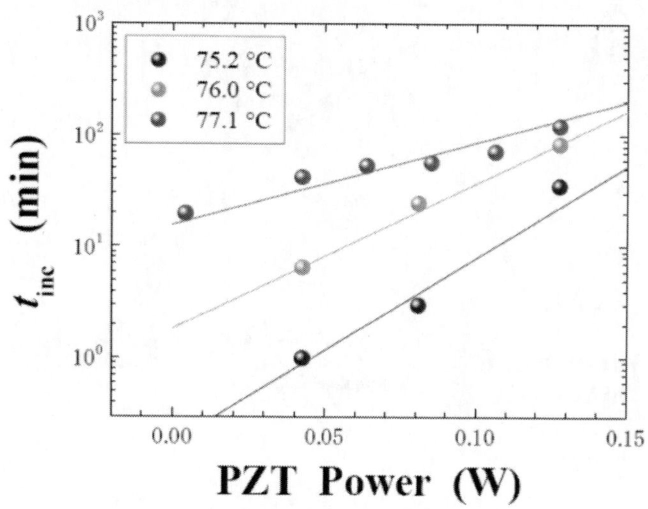

Figure 12. Transducer power (*P*) and holding temperature dependence of inculcation time in the quenched naphthalene. Solid lines are the calculated incubation time by introducing coupling term between *P* and the holding temperature.

CONCLUSION

We show the predominant anomalies of the shear modulus in metallic sodium as a monatomic system. Quite large hysteresis and anomalous *G*" in the supercooled sodium support theoretically predicted polyhedral clusters. In previous studies, it is well known that *G*" is sensitive to the dynamic heterogeneity. Mechanical shear loss described by relaxation process could play an important role not only with the supercooled liquid but also with complex disordered system.

In addition, we mentioned 'complexity' in a series of simple molecular system. Dissipative system at non-equilibrium state is schematically drawn in Figure 13(a). External ultrasonic field corresponds to vibrational energy injection into system, while quenching causes thermal dissipation from the system. In the extreme environment, a hidden character in the simple molecular system is exposed to us.

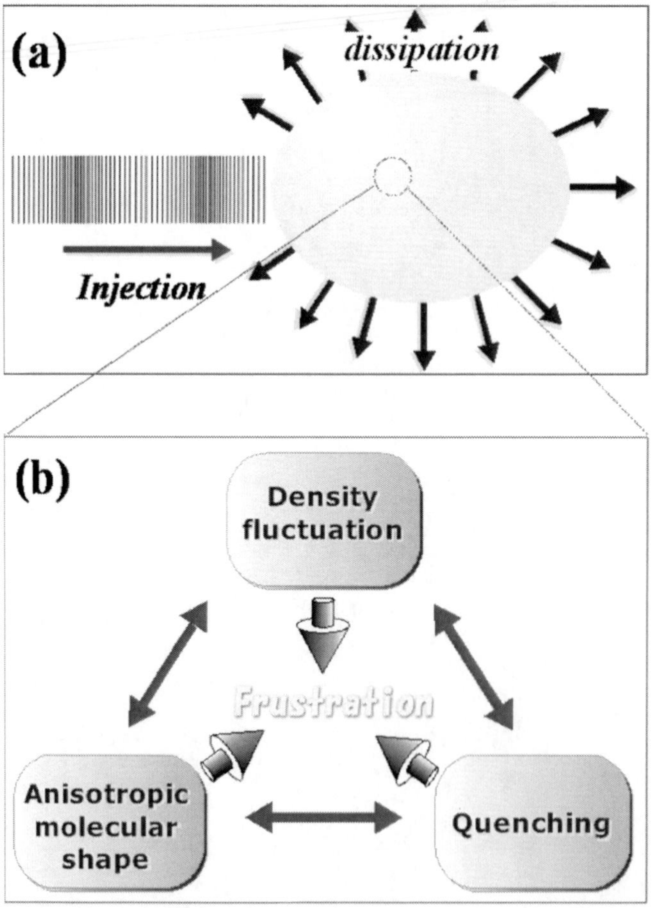

Figure 13. Dissipative system in supercooled liquid. (a) Macroscopic energy injection and thermal dissipation. (b) Microscopic factors (local density fluctuation induced by longitudinal ultrasonic irradiation, anisotropic shape factor of molecule and random orientation in quenching) and their interactions. Frustration in simple system is realized by a competition among the factors.

Microscopically, inherent frustration is induced by three factors (Figure 13(b)): (i) local spatial- and time-density fluctuation by external longitudinal ultrasonic filed, (ii) molecular packing efficiency depending on anisotropic molecular shape and (iii) random orientational order caused by quenching. A significant finding is extraordinal nucleation and growth process in the quenched benzene. In addition, comparing with normal supercooled naphthalene provides an insight for better understanding. In the quenched benzene, a new type of fluctuations is enhanced by a strong coupling with external ultrasound fields even in simple molecular system.

ACKNOWLEDGMENTS

Author appreciates helpful discussions with Professor H. Yoshizaki and Professor H. Matsumoto of National Defense Academy.

REFERENCES

[1] P. G. Debenedetti and F. H. Stillinger, *nature* 410 (2001) 259.
[2] L.-M. Martinez and C. A. Angell, *nature* 410 (2001) 663.
[3] H. Shintani and H. Tanaka, *nature phys.* 2 (2006) 200.
[4] V. Van Hoang and T. Odagaki, *J. Phys. Chem. B* 115 (2011) 6946.
[5] S. K. Lai and H. C. Chen, *J. Phys.: Condens. Matter* 5 (1993) 4325.
[6] S. K. Lai and S. Y. Chang, *Phys. Rev. B* 51 (1995) 12869.
[7] Z. Hou, R. Liu, H. Liu, Z. Tian, X. Wang, Q. Zhou and Z. Chen, *J. Chem. Phys.* 127 (2007) 174503.
[8] K. Tohji and Y. Murata, *Jpn. J. Appl. Phys.* 21 (1982) 1199.
[9] M. Misawa and T. Fukunaga, *J. Chem. Phys.* 93 (1990) 3495.
[10] H. B. Bürgi, S. C. Capelli, A. E. Goeta, J. A. K. Howard, M. A. Spackman and D. S. Yufit, *Chem. Euro. J.* 8 (2002) 3512.
[11] W. I. F. David, R. M. Ibberson, G. A. Jeffrey and J. R. Ruble, *Physica. B* 180-181 (1992) 597.
[12] D. W. J. Cruickshank, *Acta. Crystallogr.* 10 (1957) 504.
[13] G. S. Pawley and E. A. Yeats, *Acta. Crystallogr.* B 25 (1969) 2009.
[14] K. Ishii, H. Nakayama, T. Yoshida, H. Usui and K. Koyama, *Bull. Chem. Soc. Jpn.* 69 (1996) 2831.
[15] K. Ishii, H. Nakayama, M Kawahara, K. Koyama, K. Ando and J. Yokoyama, *Chem. Phys.* 199 (1995) 245.
[16] K. Ishii, H. Nakayama, Y. Yagasaki, K. Ando and M. Kawahara, *Chem. Phys. Lett.* 222 (1994) 117.
[17] W. P. Mason, W. O. Baker, H. J. McSkimin and J. H. Heiss, *Phys. Rev.* 75 (1949) 936.
[18] R. Meister, C. J. Marhoeffer, R. Sciamanda, L. Cotter and T. Litovitz, *J. Appl. Phys.* 31 (1960) 854.
[19] A. E. Clark and T. A. Litovitz, J. *Acous. Soc. Ame*. 32 (1960) 1121.
[20] R. Kono and H. Yoshizaki, *J. Chem. Phys.* 75 (1981) 4654.

[21] H. Yoshizaki, *Polymer J.* 28 (1996) 790.

[22] L. E. Kinsler, A. R. Frey, A. B. Coppens and J. V. Sanders, *Fundamentals of Acoustics*, 4th Ed. Wiley, New York, 2000.

[23] H. Abe, H. Yoshizaki, N. Ishii and A. Mine, *Memoirs Nat. Def. Acad. Jpn.* 40 (2000) 53.

[24] H. Abe, T. Sakurai and H. Yoshizaki, *J. Non-Cryst. Solids* 357 (2011) 3080.

[25] P. He and J. Zheng, *Ultrasonics* 39 (2001) 27.

[26] M. Molero, I. Segura, S. Aparicio, M. G. Hernández and M. A. G. Izquierdo, *Ultrasonics* 50 (2010) 824.

[27] H. Abe, I. Hori, T. Yatomi, M. Kikuchi, H. Matsumoto and H. Yoshizaki, *Jpn. J. Appl. Phys.* 42 (2003) 3552.

[28] S. Hansen, *Rheol. Acta.* 47 (2008) 169.

[29] H. Yoshizaki, *Polymer J.* 25 (1993) 553.

[30] C. Maggi, B. Jakobsen, T. Christensen, N. B. Olsen and J. C. Dyre, *J. Phys. Chem. B* 112 (2008) 16320.

[31] H. Abe, K. Ohshima and T. Suzuki, *Mater. Trans. JIM* 36 (1995) 1200.

[32] M. L. Cowan, I. P. Jones, J. H. Page and D. A. Weitz, *Phys. Rev. E* 65 (2002) 066605.

[33] Y. Noda, S. Nishihara and Y. Yamada, *J. Phys. Soc. Jpn.* 53 (1984) 4241.

[34] N. Hamaya, Y. Yamada, J. D. Axe, D. P. Belanger and S. M. Shapiro, *Phys. Rev. B* 33 (1986) 7770.

[35] T. Kakeshita, K. Kuroiwa, K. Shimizu, T. Ikeda, A. Yamagishi and M. Date, *Mater. Trans. JIM* 34 (1993) 415.

[36] H. Abe, M. Ishibashi, K. Ohshima, T. Suzuki, W. Wuttig and K. Kakurai, *Phys. Rev. B* 50 (1994) 9020.

[37] H. Abe and K. Ohshima, *J. Phys. Soc. Jpn.* 67 (1998) 2952.

[38] A. A. Burbelko, E. Fra´s and W. Kapturkiewicz, *Mater. Sci. Eng. A* 413–414 (2005) 429.

In: Ultrasonics: Theory, Techniques and Practical Applications ISBN: 978-1-62257-685-2
Editors: Hanako Ayabito and Mitsuko Katsukawa

Chapter 6

Improved Mathematical Model of a Traveling Wave Ultrasonic Motor

S. Belkhiat, Z. Boumous and F. Z. Kebbab
Laboratoire DAC HR Université Sétif I, Algeria

Abstract

The traveling wave ultrasonic motor has excellent characteristics. It has advanced features such as high-holding torque without supply high torque at low speed, absence of parasitic magnetic fields, continuous contact, high response, quiet operation and small size. All these properties make of this motor an ideal actuator, comparatively with the electromagnetic one. During the last decade, many prototypes of ultrasonic motor have been designed as a function needs. The advanced features are due to the working principle that is based on high-frequency vibrations and due to the tribological phenomenon acting in the zone of contact between stator and rotor. There is not a perfect contact model that can control the contact dynamic between the rotor and the stator because of it has a nonlinear behavior. For this reason, almost all research carried out is based on a requirement that leads to some errors between the experimental and theoretical results. Modeling of traveling wave ultrasonic motors (TWUM), has been, so, subject much research.

In this chapter, optimal deformation angle and resonance frequency of frictional material are investigated. Force expressions in the stator/rotor interface are developed. Mechanical characteristics taking into account the effect of shearing deformation are deduced and presented. Results of the simulation are validated using the measurements made by other researchers on traveling wave piezoelectric motor AWM90-X, Daimler-Benz type, and a prototype USM (AWM-90).

I. Introduction

The first scientific document concerning piezoelectric motors has been applied in 1942. It proposes a motor that is based on piezoelectric excited vibrations which should be transformed to rotary motion by gear wheel mechanism. After progresses in material science and power electronics first rotary and later the linear ones have been developed in the 1970s. Their operating principle is based on piezoelectric actuators that convert electric energy to mechanical energy in the form of vibrations of an elastic body [1- 2]. This body is excited by the inverse piezoelectric effect in such a manner that points of its surface perform elliptic motions with a frequency in the ultrasonic range. Another body is pressed beside the vibrating body by a static normal force F_N and thus the microscopic small vibrations are transformed to continuous motion by friction forces [3]. Based on this operating principle many different types of ultrasonic piezoelectric motors have been developed. Their main advantages over the conventional electromagnetic devices are:

- Different velocities without gear-mechanisms,
- High positioning accuracy due to the friction drive,
- High holding torque (braking force without energy supply) [4],
- Simplicity and flexibility in structural design [4 -5],
- No magnetic noise [6],
- High out put torque at low speed [7],
- High force density.

These properties make ultrasonic motors suitable to a direct-drive mechanism. Today ultrasonic piezoelectric motors are used in applications where their characteristic advantages balance their costs, which are still too high for high volume applications, like robotic, medical, space and automotive applications [8]. For that, generally, various ultrasonic motors are designed as a function needs. Multilayer piezoelectric actuators, for example, are now commonly used in some industrial applications. The more significative is the Piezoelectric Common Rail Injectors for diesel engine such as the Siemens AG or Bosch solutions. Petit et al. [9-12], used in theirs ultrasonic motor prototype a functional traveling wave induced by longitudinal actuators (*TWILAs*) motor which combines good mechanical performances and low production cost. This motor belongs to the traveling wave ultrasonic motor (TWUM) family because it uses the propagation of a flexional traveling wave inside a ring-shaped stator. However, to overcome some limitations, a new version of the TWILA Integrated Multi Layer actuators (so-called ML TWILA) [12] has been designed, by the same author, with two purposes:

- to decrease the supply voltage
- to open new perspectives of easy-strategy command.

Up to now, the analyses of friction of ultrasonic motors and several conclusions based on tests and experiments were carried out [13]. There is not a perfect contact model that can control the contact dynamic between the rotor and the stator because of it has a nonlinear behavior. For this reason, almost all research carried out is based on a requirement that leads to some errors between the experimental and theoretical results [14]. For example, Kurosawa and Ueha in [15] in 1988 developed a model that consists of a rotor and a stator that are viewed as rigid bodies. They calculated the mechanical characteristics of the motor, but some details are not considered, such as the shearing deformation of the contact layer. Nevertheless, they did not analytically explain this phenomenon. In this way, a model of stator/rotor contact has been developed [16] in order to predict the motor behavior. The paper pointed out the importance of the tangential elasticity of the contact layer, which was compared with a rigid contact layer.

During the last five years, a series of published papers has considered the shearing deformation [15–17]. Different studies have been carried out. The electromechanical responses of the piezoelectric actuator in harmonic vibration have been calculated and analyzed by the three-dimensional FEM [17–18]. The motor's mechanical characteristics have been deduced and discussed, but the transient response has not studied. Other modeling methods based on equivalent circuit methods, analytical approaches and finite element modeling are the most common methods used. The finite element modeling method provides the possibility of a precise study on different mechanical parts of the motor, but the electrical parameters are not considered. The complexity and computational requirements of FEM models as a design tool [17] make them improper for practical integration in the overall motor model [17]. In these methods, fewer interface parameters are considered, so they are not suitable for analysis of parameters and optimizing the design of a system [15]. Likewise, electrical parameters are not considered in the FEM method.

In this CHAPTER, a MATLAB/SIMULINK environment is used with the goal of making a simple model that runs on a personal computer. A mathematical description of the contact mechanism is developed. All the forces acting in the interface rotor/stator are introduced in the analytical model. The forces' expressions considering the shearing deformation of frictional material are developed. The experimental shearing angle expressed in [13] introduced in the model, and the optimal value is tracked with an iteration algorithm using the MATLAB computational language. In order to validate the model, the simulation results are compared with results measured by Kandare and Wallaschek [19] on the Daimler-Benz AWM90-X motor using the test bench and are compared with the ones measured by Thomas Schulte and Norbert Frohleke [20] on a prototype USM(AWM-90).

The chapter is organized as follows: Section II presents working principle and experimental setup. The theoretical study and force calculation are synthesized in the section III. The section IV presents the simulation results and theirs discussion.

II. Working Principle and Experiment Setup

The ultrasonic motor is a new type of driver. It performs a two-step energy conversion. In the first stage, electrical energy is converted into mechanical vibrations of the stator. In the second stage, high-frequency oscillatory vibrations of the stator are transformed into unidirectional microscopic motion of the rotor. Several references concerning the working principle of piezoelectric ultrasonic motors are given in [19].

Several types of piezoelectric ultrasonic motors have been suggested and designed in the last 25 years. The large family of USMs can be categorized by various criteria: based on their functionality into two types, linear and rotary motor, also, by the exploitation of the piezoelectric material to adjust the propagation of its vibration wave in the static part of the USM motors (stator). With this classification, standing wave and traveling wave USMs can be identified. Following the advantages and various applications of rotary type ultrasonic motors, and in particularly the traveling wave ones, compared to the other types of USMs, have attracted more research attention to them. Figure 1 shows an exploded view of a typical traveling wave ultrasonic motor, is discussed in this chapter.

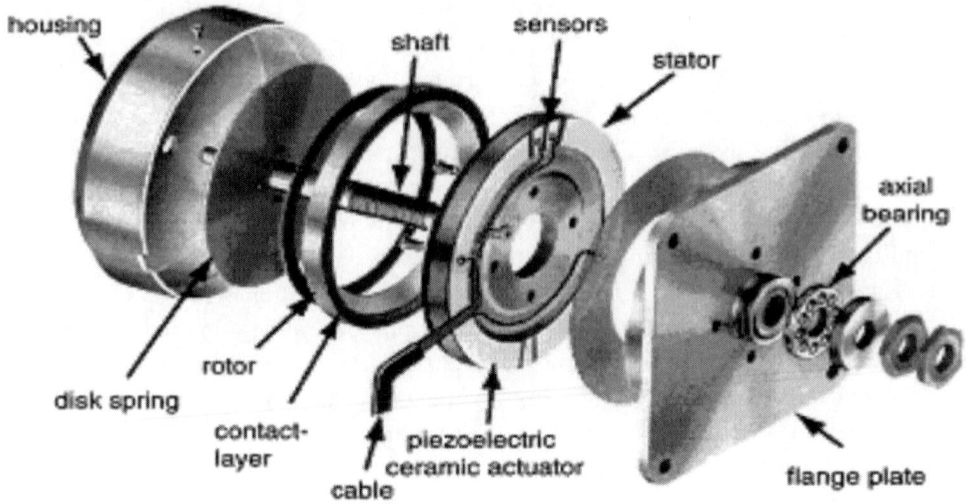

Figure 1. An exploded view of a typical traveling wave ultrasonic motor.

It consists of two basic parts: the statically part vibration (stator vibration) with a frequency in the ultrasonic range, and the driven part (rotor) by the stator effect via frictional forces. Stator is composed of an elastic body and a thin piezo-ceramic ring. The piezo-ceramic ring is bonded under the elastic body. It has the function of exciting traveling bending waves and is shown in Figure 2.

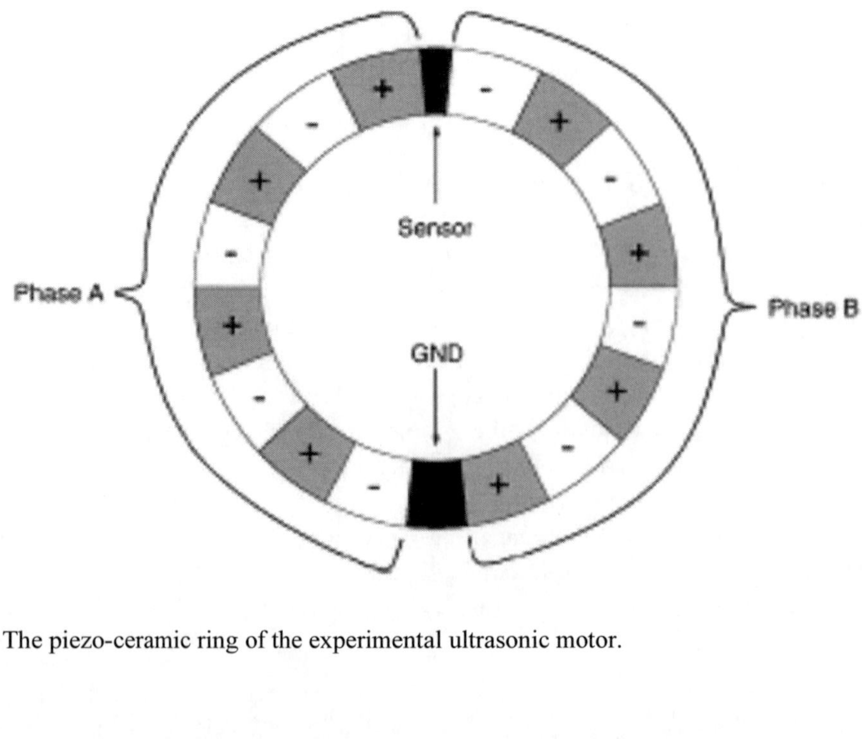

Figure 2. The piezo-ceramic ring of the experimental ultrasonic motor.

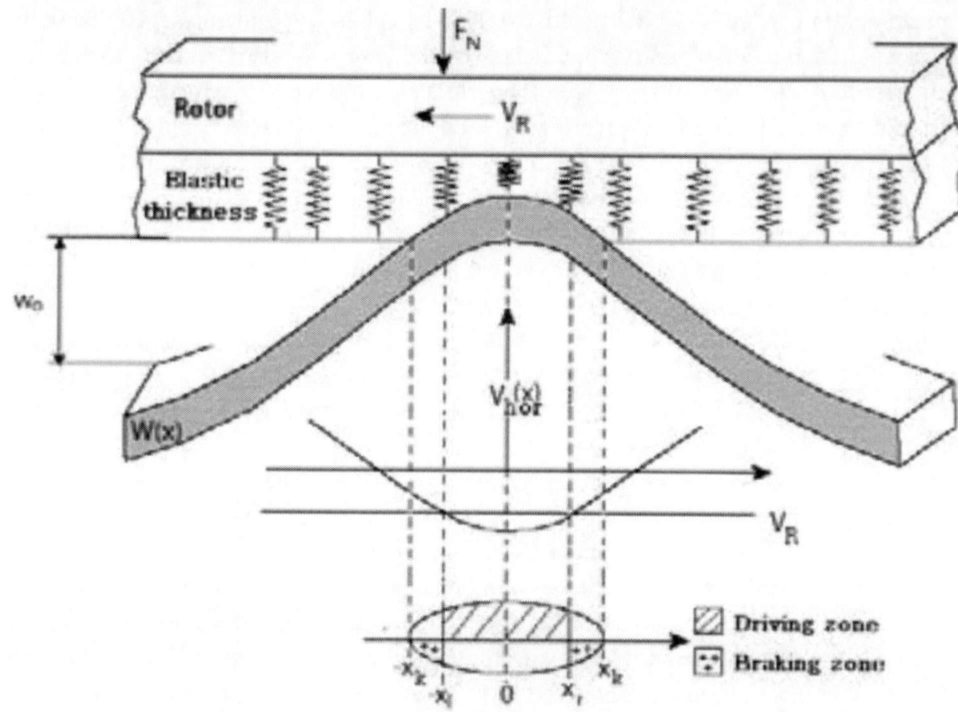

Figure 3. Drive of the traveling wave piezoelectric motor by the force of friction.

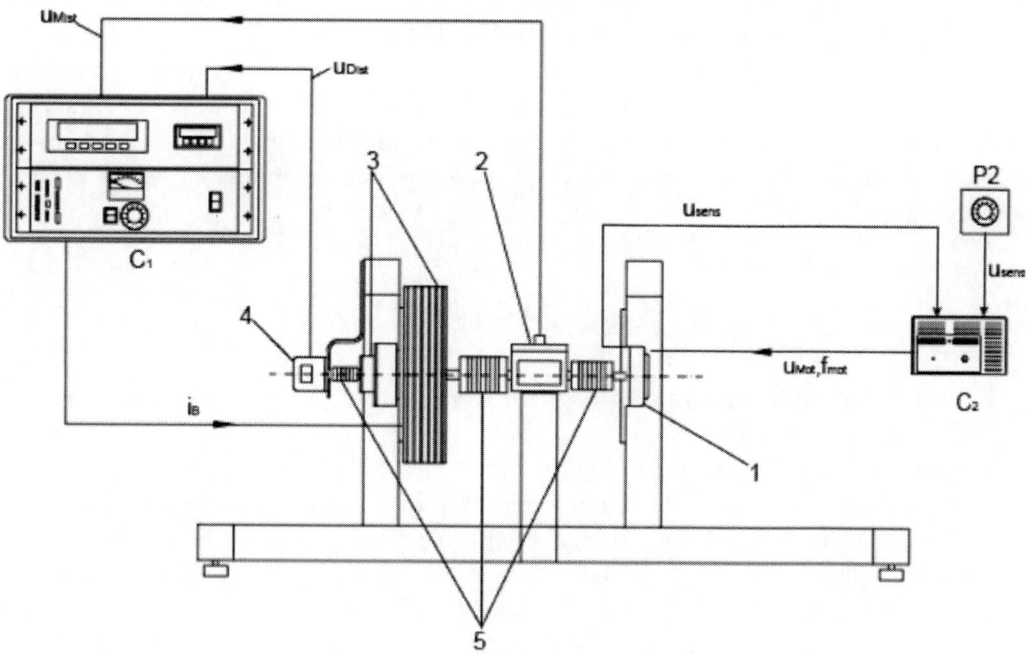

Figure 4. Working principle and experiment system of ultrasonic motor : (a) Test bench for measurement of speed-torque curves of traveling wave motors (1) Traveling wave; (2) Torque sensor; (3) Magnetic powder brakes; (4) Incremental optical encoder; (5) Metal clutch.[24].

The piezo-ceramic ring is divided into two halves: phase A and phase B. These two phases are separated by sensor and ground parts which are a quarter and 3 quarters of a wavelength, respectively. Each phase (A or B) includes n segments. Each segment is a half wavelength and polarized adversely regarding the adjacent one. Phase A and phase B are a quarter of the wavelength out of phase, spatially. The phases are excited by two sinusoidal voltages which are temporally 900 out of phase. More details are given in [21-23] Therefore, a traveling wave is generated and the particles of the stator surface move elliptically. The sensor part is used for measuring the amplitude and the phase of the traveling wave to control the excitation of the piezo-ceramic ring.

The rotor is pressed against the stator by means of a disk spring, and a thin contact layer is bonded to the rotor in the contact region. Therefore, the vibration of the stator with high frequency and small amplitude is transformed into the macroscopic rotary motion of the rotor by friction.

In a conventional type of TWUM, the traveling wave is generated by superimposing two orthogonal standing waves on the stator ring. A piezoelectric ceramic is bonded to one side of the stator, and the traveling wave is excited in the stator by applied electrical signals on the piezoelectric ceramic. The rotor is in contact with the stator at the crests of the traveling wave and is pressed against the stator with a normal force. As the traveling wave propagates on the stator, points on the stator surface undergo an elliptical motion. The tangential displacement of the stator at the contact interface drives the rotor by means of the friction force (Figure 3). The working principle and test bench for the measurements are shown in Figure 4.

With the potentiometer on the control unit C_1, load torque is set and compared to the actual torque (U_{Mist}) obtained from the torque sensor. From the difference, the PI controller sets the appropriate input current (i_B) for the brake. On the control unit C_1, there are also two displays where load torque and rotational speed can be observed. Control of the sensor voltages is realized with control unit C_2. With the potentiometer P_2, the desired values of sensor voltages are set, and controller C_2 compares this value with that one obtained from the piezo-ceramic sensor of the motor (U_{sens}). The corresponding frequency is set so as to minimize the difference.

III. Forces Calculation

III.1. Force Calculation not Taking into Account Shearing Deformation

The rotor is pressed by means of the force F_N (Figure 3). Frictional forces appear in the rotor/stator contact and transmit a traction force to the rotor. F_N also has the role of fixing the holding torque without supplying the stator with electrical energy. The traction effect is accompanied with the effects of feedback, which are caused by the axial displacement of stator and reaction of the interface (rotor/stator). In this work, Coulomb friction is assumed. We account for feedback forces and shearing deformation in the resulting frictional force. Two working zones are considered: the driving zone and the breaking zone, such as represented in Figure 4. In the contact layer there is a normal pressure between the stator and the rotor along the contact zone [13]. The line force distribution along the contact region between stator and rotor is given by

$$f_Z = \frac{E.b}{h}\Delta W = C_N.\Delta W \tag{1}$$

$$\Delta W = (\cos(k\widetilde{x}) - \cos(kx_k)).W_0 \tag{2}$$

The superposition of the two standing waves generates the traveling wave that propagates in the stator ring. In mathematical terms, this is expressed by

$$W(x, t) = W_1(x, t) + W_2(x, t) = W_{1m}(t)\sin(kx) + W_{2m}(t)\cos(kx) \tag{3}$$

$$W_{1m}(t) = W_0 \sin(\omega t) \tag{4}$$

$$W_{2m}(t) = W_0 \cos(\omega t) \tag{5}$$

where W_{1m} and W_{2m} are the modal displacements.

After some mathematical simplifications we get the expression of the vertical displacement of the stator.

$$W(\tilde{x}) = (W_{02}((\cos^2(\omega t) + \sin^2(\omega t))^{1/2} \cos(k\tilde{x}) = W_0 \cos(k\tilde{x}) \tag{6}$$

where W_0 is the amplitude of the traveling wave and $\tilde{x}$ is the new coordinate system moving with traveling wave. CN and h are the rigidity and the thickness of the contact layer, respectively. b is the width of the contact in the axial direction, and E is the Young's modulus of the contact material.

The normal force acting on the rotor is expressed by,

$$F_N = n\int_{-X_k}^{X_k} f_z(\widetilde{x})d\widetilde{x} \tag{7}$$

$$\tilde{x} = x - \frac{\omega}{k}t \tag{8}$$

where n is the number of wave crests and $\widetilde{x}$ is the new coordinate system moving with the travelling wave.

The force that drives the rotor can be calculated as follows:

$$F_{ant} = n\mu\int_{-X_k}^{X_k} \sin g(v_{hor}(\widetilde{x}) - v_R) f_z(\widetilde{x})d\widetilde{x} \tag{9}$$

where v_R is the rotor speed, v_{hor} is the velocity of the stator surface points in the horizontal direction and μ is the friction coefficient.

$$\sin g(v_{hor}(\tilde{x}) - v_R) = \begin{cases} 1 & v_{hor}(\tilde{x}) \rangle v_R \\ 0 & v_{hor}(\tilde{x}) = v_R \\ -1 & v_{hor}(\tilde{x}) \langle v_R \end{cases} \tag{10}$$

$$F_{ant} = n\mu C_N W_0 \int_{-X_k}^{X_k} \text{sing}(v_{hor}(\tilde{x}) - v_R).(\cos(k\tilde{x}) - \cos(kx_k))d\tilde{x} \tag{11}$$

$$F_{ant} = n\mu C_N W_0 \left[- \int_{-X_k}^{-X_{sl}} f_Z(\tilde{x})d\tilde{x} + \int_{-X_r}^{X_{sr}} f_Z(\tilde{x})d\tilde{x} - \int_{X_{sr}}^{X_k} f_Z(\tilde{x})d\tilde{x} \right] \tag{12}$$

$$F_{ant} = \frac{n\mu C_N W_0}{K} \begin{bmatrix} 2(-\sin(kx_k) + kx_k \cos(kx_k)) + \\ 2(\sin(kx_{sr}) - kx_{sr} \cos(kx_k)) + \\ (\sin(kx_{sl}) - kx_{sl} \cos(kx_k)) + \\ (\sin(kx_r) - kx_r \cos(kx_k)) \end{bmatrix} \tag{13}$$

$$F_{ant} = \frac{n\mu C_N W_0}{K} \begin{bmatrix} -2.\xi(x_k) + 2.\xi(x_{sr}) + \\ \xi(x_{sl}) + \xi(x_r) \end{bmatrix} \tag{14}$$

$$F_{ant} = \frac{n\mu C_N W_0}{K} \begin{bmatrix} -2.\xi(x_k) + 2.\xi(x_{sr}) + \\ \xi(x_{sl}) + \xi(x_r) \end{bmatrix} \tag{15}$$

The feedback effects of the interface (rotor/stator) are characterized by the feedback reaction in the tangential direction. The tangential force is calculated as follows:

$$F_{Tnorm} = -n \int_{-X_k}^{X_k} \Phi_x^T f_z(\tilde{x})d\tilde{x} \tag{16}$$

$$\Phi_x^T = [\sin(kx) \quad \cos(kx)] \tag{17}$$

$$F_{TF} = -n\mu a \int_{-X_k}^{X_k} \frac{\partial \Phi_x^T}{\partial x} \sin g(v_{hor}(\tilde{x}) - v_R).f_Z(\tilde{x})d\tilde{x} \tag{18}$$

$FT = n\mu aCN\ [-E_c(-x_{sl}) - E_2(x_k) + E_c(x_{sr})]$

$$E_C(\tilde{x}) = kW_0 \int \begin{bmatrix} \cos(KX) \\ -\sin(KX) \end{bmatrix} .p(\tilde{x})d\tilde{x}$$

$$E_C(x) = \begin{bmatrix} W_{1m}\tau_2 + W_{2m}\tau_1 \\ W_{2m}\tau_2 - W_{1m}\tau_1 \end{bmatrix} = [E_1 + E_2] \tag{19}$$

$$\tau_1(\tilde{x}) = \frac{1}{2}\sin^2(k\tilde{x}) - \cos(kx_k)\cos(k\tilde{x})$$
$$\tau_2(\tilde{x}) = \frac{k\tilde{x}}{2} + \frac{1}{4}\sin(2k\tilde{x}) - \cos(kx_k)\sin(k\tilde{x}) \quad (20)$$

III.2. Forces Calculation Taking into Account Shearing Deformation

The improved model, taking account the shearing deformation is developed in [25]. The shearing deformation zone, the driving zone and the breaking zone are shown in Figure 5.

The contact model between the rotor and the stator that takes into account the shearing deformation is shown in Figure 6.

Since the frictional material is softer than the stator, it produces shearing deformation in the contact zone where the sticking region is [13, 25]

$$\Delta x = h \tan\theta \quad (21)$$

where θ is the shearing angle.

Eq. (13) shows the force that drives the rotor, where *xk* is contact point, x_r is the no-slip point shearing deformation considered, x_{sl} is the left no-slip point, and x_{sr} is the right no-slip point.

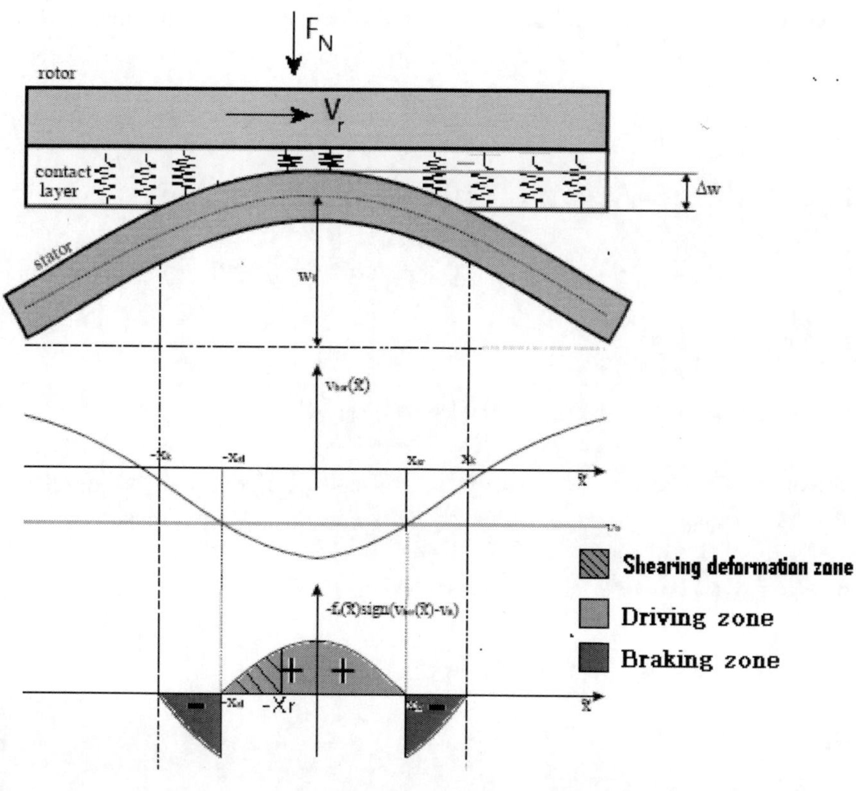

Figure 5. Case 1 $X_{sl} \le X_K$ *and* $X_{sr} \le X_K$.

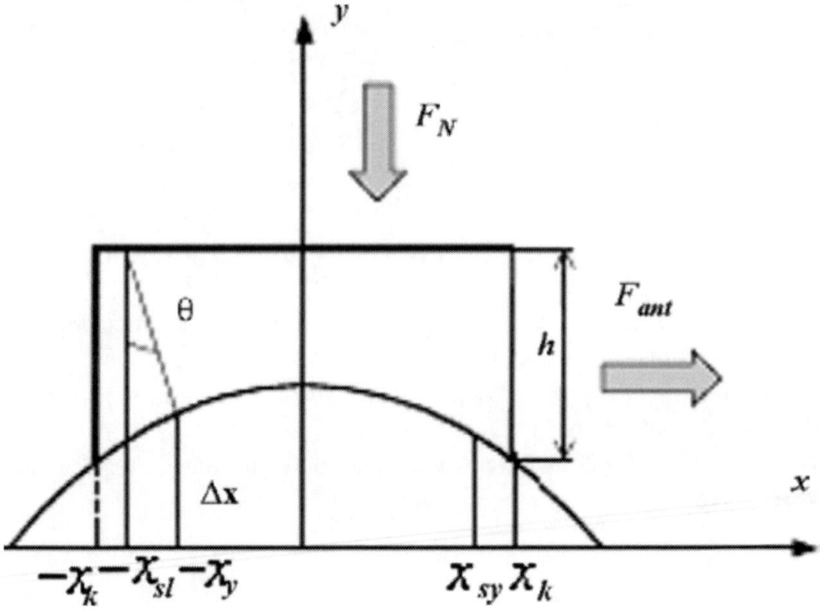

Figure 6. The contact model between the rotor and the stator, taking account of shearing de formation.

The stator sticks the rotor in the region [$-x_{sl}$, $-x_r$] and slides again after $-x_r$. In order to calculate the traction and the feedback forces, one must initially determine the no-slip points and the length of the contact zone. The no-slip points can be obtained using the equality between the rotor speed and the horizontal speed ofthe stator surface points. The phase difference (ϕ_c) is shown in Figure 7.The points x_k, x_r, x_{sl} and x_{sr} are calculated as follows:

$$v_R = v_0(t)\cos(k\,x + \phi_c(t)) \tag{22}$$

$$kx_{sr} + \phi_c(t) = a\,(\cos(v_R/v_0(t))x_{sr} = (1/k)\,(a\cos(v_R/v_0(t) - \phi c(t)) \tag{23}$$

$$-(-x_{sr} + (1/k)\phi_c(t)) = x_{sl} + (1/k)\,\phi_c\,(t)\;x_{sl} = x_{sr} + (2/k)\phi_c(t)$$

Finally,

$$kx_{sr} + \phi_c(t) = (1/k)\,(a\cos\,(v_R/v_0(t)) + \phi_c(t)) \tag{24}$$

When the vertical displacement of the rotor, $W_{rot}(t)$, is equal to the vertical displacement of the stator surface at the edge of the contact zone, $W_{rot}(t)$ is expressed by

$$w_{rot}\,(t) = w_0(t)\cos(kx_k) \tag{25}$$

x_k can be calculated from formula (26):

$$X_k = \frac{1}{k}acos(\frac{W_{rot}(t)}{W_0(t)}) \tag{26}$$

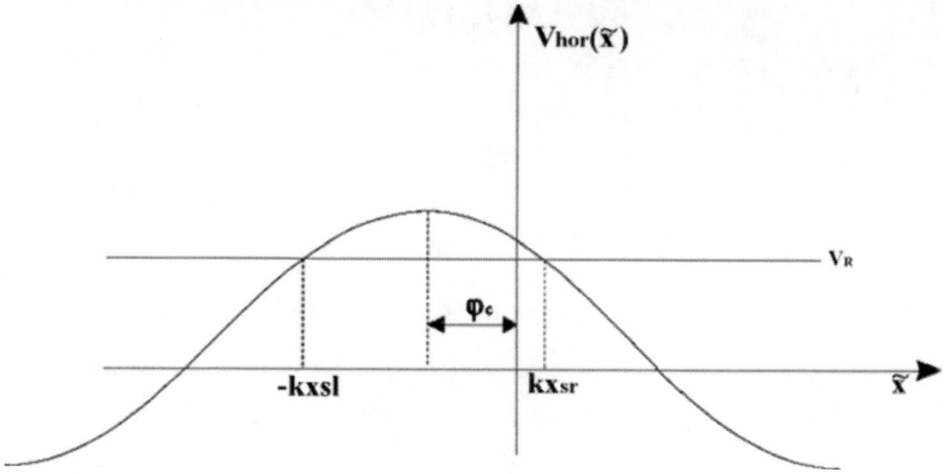

Figure 7. Position of no-slip points according to phase difference ϕc.

If the position of the rotor is higher than the positive crest of the displacement wave ($w_{rot}(t) > w_0(t)$), then there is no contact between the stator and the rotor, and $x_k = 0$.

If $w_{rot}(t)$ is lower than the negative point ($w_{rot}(t) < -w_0(t)$), then total penetration of stator occurs and $x_k = \pi/k$.

$$X_k = \begin{cases} \frac{1}{k}\operatorname{acos}\left(\frac{W_{rot}(t)}{W_0(t)}\right), |W_{rot}(t) \leq |W_0(t)|| \\ 0, W_{rot}(t) > W_0(t) \\ \frac{\pi}{k}, W_{rot}(t) \langle -W_0(t) \end{cases} \tag{27}$$

Finally, one can deduce x_r taking into account the shearing deformation.

$$x_r = x_{sl} - \Delta x \tag{28}$$

$$F_{ant} = n\mu \int_{-X_k}^{X_k} \sin g(v_{hor}(\tilde{x}) - v_R) f_z(\tilde{x}) d\tilde{x} \tag{29}$$

where v_R is the rotor speed, v_{hor} is the velocity of the stator surface points in the horizontal direction and μ is the friction coefficient.

$$\sin g(v_{hor}(\tilde{x}) - v_R) = \begin{cases} 1 & v_{hor}(\tilde{x}) \rangle v_R \\ 0 & v_{hor}(\tilde{x}) = v_R \\ -1 & v_{hor}(\tilde{x}) \langle v_R \end{cases} \tag{30}$$

$$F_{ant} = n\mu C_N W_0 \int_{-X_k}^{X_k} \operatorname{sing}(v_{hor}(\tilde{x}) - v_R).(\cos(k\tilde{x}) - \cos(kx_k)) d\tilde{x} \tag{31}$$

$$F_{ant} = n\mu C_N W_0 \left[-\int_{-X_k}^{-X_{sl}} f_Z(\tilde{x}) d\tilde{x} + \int_{-X_r}^{X_{sr}} f_Z(\tilde{x}) d\tilde{x} - \int_{X_{sr}}^{X_k} f_Z(\tilde{x}) d\tilde{x} \right] \tag{32}$$

$$F_{ant} = \frac{n\mu C_N W_0}{K}\begin{bmatrix} 2(-\sin(kx_k) + kx_k \cos(kx_k)) + \\ 2(\sin(kx_{sr}) - kx_{sr} \cos(kx_k)) + \\ (\sin(kx_{sl}) - kx_{sl} \cos(kx_k)) + \\ (\sin(kx_r) - kx_r \cos(kx_k)) \end{bmatrix} \tag{33}$$

$$F_{ant} = \frac{n\mu C_N W_0}{K}\begin{bmatrix} -2.\xi(x_k) + 2.\xi(x_{sr}) + \\ \xi(x_{sl}) + \xi(x_r) \end{bmatrix} \tag{34}$$

The feedback effects of the interface (rotor/stator) are characterized by the feedback reaction in the tangential direction. The tangential force is calculated as follows:

$$F_{Tnorm} = -n\int_{-X_k}^{X_k} \Phi_x^T f_z(\tilde{x})d\tilde{x} \tag{35}$$

$$\Phi_x^T = [\sin(kx) \quad \cos(kx)] \tag{36}$$

$$F_{TF} = -n\mu a \int_{-X_k}^{X_k} \frac{\partial \Phi_x^T}{\partial x} \sin g(v_{hor}(\tilde{x}) - v_R).f_Z(\tilde{x})d\tilde{x} \tag{37}$$

$$F_{TF} = n\mu a C_N \begin{bmatrix} -E_c(-x_{sl}) + E_c(-x_k) + E_c(x_{sr}) - \\ E_c(-x_r) - E_c(x_k) + E_c(x_{sr}) \end{bmatrix} \tag{38}$$

$$F_{TF} = -n\mu a C_N \begin{bmatrix} -E_c(-x_{sl}) - E_c(-x_r) - \\ 2E_2(x_k) + 2E_c(x_{sr}) \end{bmatrix} \tag{39}$$

$$E_C(\tilde{x}) = kW_0 \int \begin{bmatrix} \cos(KX) \\ -\sin(KX) \end{bmatrix}.p(\tilde{x})d\tilde{x}$$
$$= \begin{bmatrix} W_{1m}\tau_2 + W_{2m}\tau_1 \\ W_{2m}\tau_2 - W_{1m}\tau_1 \end{bmatrix} = [E_1 + E_2] \tag{40}$$

$$\tau_1(\tilde{x}) = \frac{1}{2}\sin^2(k\tilde{x}) - \cos(kx_k)\cos(k\tilde{x})$$
$$\tau_2(\tilde{x}) = \frac{k\tilde{x}}{2} + \frac{1}{4}\sin(2k\tilde{x}) - \cos(kx_k)\sin(k\tilde{x}) \tag{41}$$

The simulation results of mathematical model are given and discussed in the section IV.

IV. SIMULATION RESULTS

The mathematical model for TWUM was used as a base [19]. Optimal frequency was determined and evaluated up to 46.65 kHz. The excitation voltage amplitude and the phase difference are 570V and /2-rd, respectively. The other parameters used for the AWM90-XDaimler-Benz motor are illustrated in Table 2. The model was run with various shearing angles. An iteration algorithm permitted us to track the optimal value of the shearing angle. The rotor speeds obtained are compared with the one measured in [19]. The results are illustrated in Table 1. The deformation angle (3°) corresponding to the speed of 45.4rpm is closer to the one measured on an unloaded AWM 90-X motor [19]. Generally, θ is very small [13], so the following simulations were carried out using on this value.

The simulation parameters used for the AWM90-XDaimler-Benz motor are given in Table 2.

The modal displacement of the stator is shown in Figure 8 for the case of vanishing load torque before and after to take into account the deformation shearing. We can see that the two modulated waves (Figure 8a and Figure 8b) are different from the point of view form and amplitude. The modulated wave, of the Figure 8a, appears non-ideal. This one takes not perfectly the form of the envelope of the modal displacements such as we see in the Figure 8b. The modulated wave, of the Figure 8a, appears non-ideal. This one takes not perfectly the form of the envelope of the modal displacements such as we see in the Figure 8b. Initially, the modal displacements of the stator are small and they increase with time for to reach a maximum of 0,9 μm at time $0{,}5\text{x}10^{-3}$s and settle to a constant value (0,6 μm) from a time of $2{,}5\text{x}10^{-3}$s.

The simulation results concerning the rotor speed for various loads, taken before and after to take into account the deformation shearing, are represented in Figures 9 and 10, respectively. The comparison of transient responses between the two cases showed a clear difference in the transient state as well as in the settling state.

Before introduction of the shearing deformation in the model, the transient state (Figure 9(a)) lasts more than 3×10^{-3} s. The oscillation's amplitude was more than 5Nm crest to crest in the starting phase. The settling state began at 3.5×10^{-3} s. However, the slip effect in the stator/rotor interface appears to start with a load of 4.3 Nm. The motor-torque characteristic quickly decreases at 4.3×10^{-3} s (see Figure 9(b)). The slip effect is not visible as a function of the rotor speed (Figure 9(c)).

Table 1. Rotor speed corresponding to angle of deformation

Angle of deformation [°]	Simulated rotor speed [rpm]
0.1	45.5
3	45.4
5	45.3
7	45.2
8	45.14
10	45.1
20	44.5
30	44.4

Table 2. Parameters of AMW 90-X motor [26]

Name	Symbol and value
Input Resistances	$Rp_1 = 5\ \Omega$ $Rp_2 = 5\Omega$
Ceramics capacity	$Cp_1 = 7.8e\text{-}^9\ F$ $Cp_2 = 7.87e\text{-}^9\ F$
Capacity of stator	$Cp_{s1} = 0.42e\text{-}^9$ $Cp_{s2} = 0.428e\text{-}^9$
Rotor Inertia	$J_R = 3.4367e\text{-}^{004}\ Kgm^2$
Radius	$R_w = 40.5e^{-3}\ m$
Effective mass	$m_{eff} = 40.5\ Kg$
Mass of rotor	$m_R = (m_{eff}+22.8+3)*1e^{-3}\ Kg$
Rotor Rigidity	$c_R = 300e^3\ N/m$
Rotor Attenuation	$d_R = 50*1e^3\ Ns/m$
Friction Coefficient	$\mu = 0.21$
Distance between the surface points of stator	$a = 4.5e^{-3}\ m$
Number of Wave Crests	$n = 11$
Resonance Frequency	$W_{res2} = w_{res1}$ $wres1 = 2*\pi*43.365*1e^3$ Hz
Wavelength	$\lambda = 2*\pi*R_w/n\ m$
Wave numbers	$k = 2*\pi/\lambda$
Contact Zone Rigidity	$c_N = 8500*1e^6\ N/m^2$
Antiresonance Frequency	$w_{ant2} = w_{ant1}$ $w_{ant1} = 2*\pi*43.425e3$ Hz
Modal mass of Stator	$m = 0.082\ Kg$
Transfer ratios	$A1 = (m*C_{p1}*(w_{ant1})^2\text{-}w_{res1})^2))^{1/2}(kgFs^{-2})^{1/2}$ $A2 = (m*C_{p2}*(w_{ant2})^2\text{-}w_{res2})^2))^{1/2}(kgFs^{-2})^{1/2}$
Stator Rigidity	$c_{S1} = (w_{res1})^2*m\ N/m$ $c_{S2} = (w_{res2})^2*m\ N/m$
Disturbance Factor	$\varepsilon_1 = \varepsilon_2, \varepsilon_1 = 0.02$
Damping ratios	$d_{S2} = d_{S1}, d_{S1} = 10\ Ns/m$

Therefore, when shearing deformation is introduced in the model (Figure 10), the settling time of the motor is reduced to 2×10^{-3} s, and the oscillation's amplitude decreases. The steady state of the motor, which occurs between 3×10^{-3} and 4.3×10^{-3} s, is characterized by linear transient responses for loads of 1.2Nm, 2.25Nm and 3.25 Nm, in comparison with the loads in Figure 9. The settling state disappears at $3.5\times10{-}3$ s. After this time, the speed decreases quickly towards zero at 4×10^{-3} s (Figure 10(b)) and towards −30rpm at 4.3×10^{-3} s (Figure 10(c)). The motor should have a fast response time and the best steady state response in this motor-torque range.

The difference between the motor-torque characteristics presented in Figures 9 and 11 clearly shows the effect of the shearing deformation. At the settling state (Figure 9; from 3×10^{-3} to 4.3×10^{-3} s), the characteristics are less linear. In the starting phase, the oscillation's amplitude changes only from a little less than 3.8 to −1.7. The overload effect (4.3Nm) of the rotor is characterized by a slip in the rotor/stator interface (Figures 10 and 11). The slip phenomenon due to an overload is clearly observed in this case. The simulation results show the dynamic workings of the motor. When the load torque increases, the rotor speed decreases.

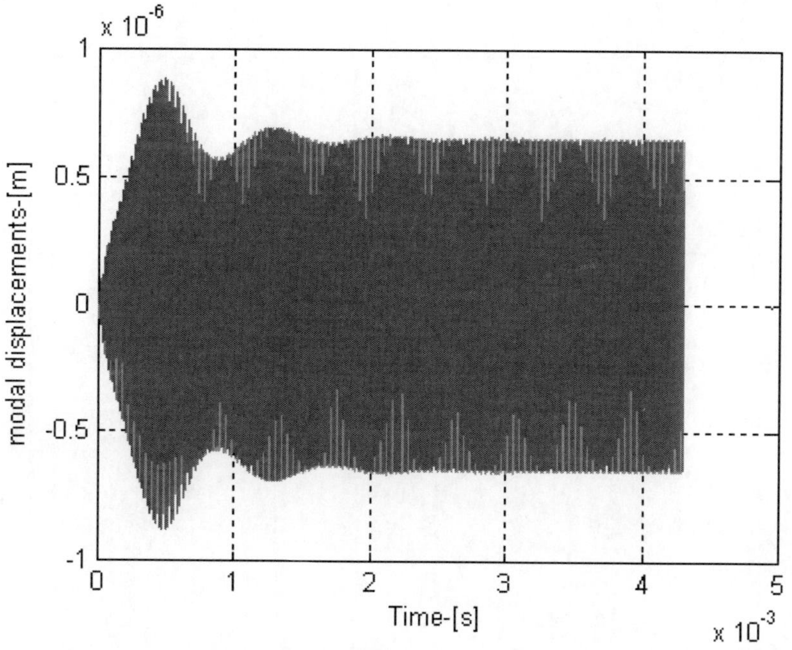

Figure 8(a). Modal displacements without shearing deformation without load.

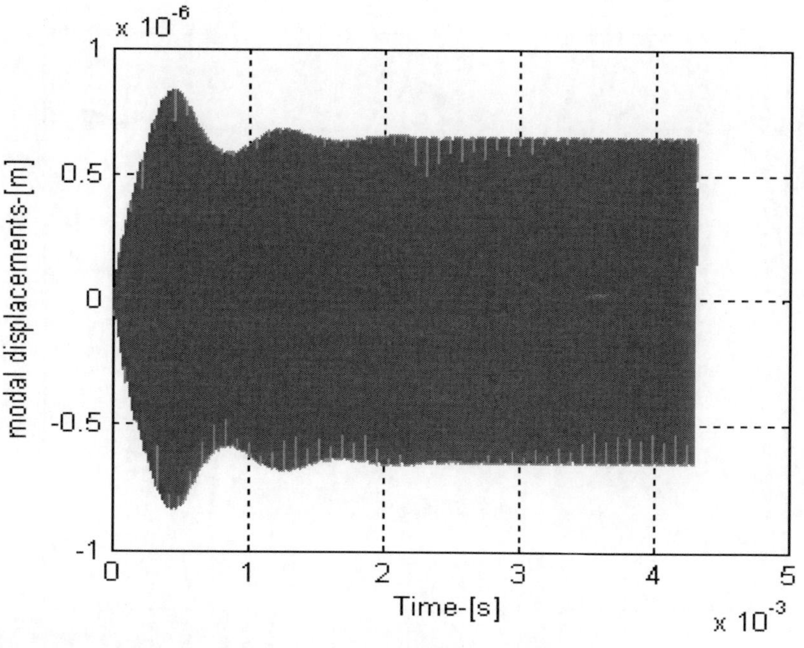

Figure 8(b). Modal displacement with shearing deformation without load.

However, the effect of shearing deformation is more apparent in the motor-torque characteristics (Figure 11). In the starting phase and for big loads more than 3.5Nm (Figure 11(b)), the oscillations are more resolved in amplitude and in breadth. The driving force is not enough to drive the overloaded rotor. It shows the rotor being broken by the load.

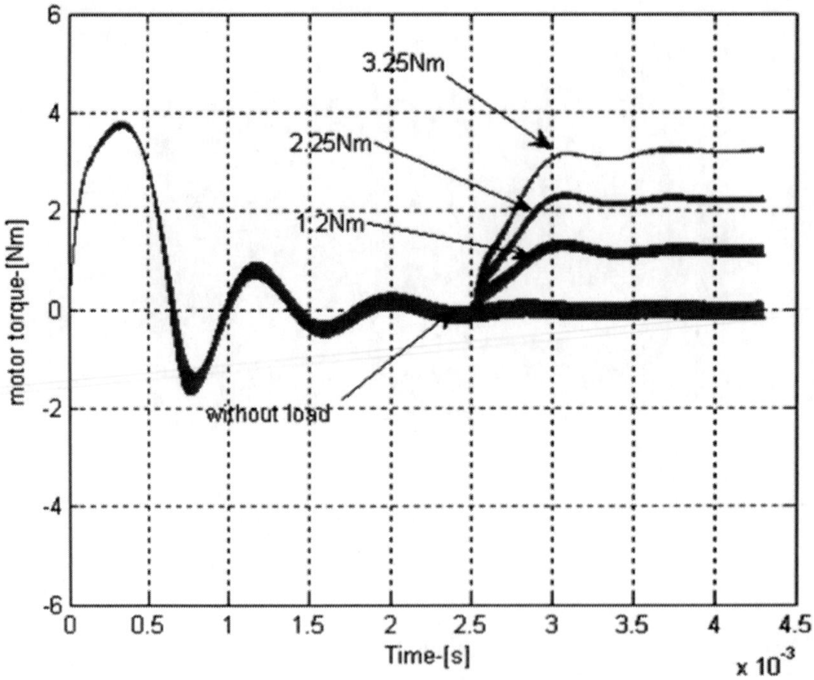

(a)

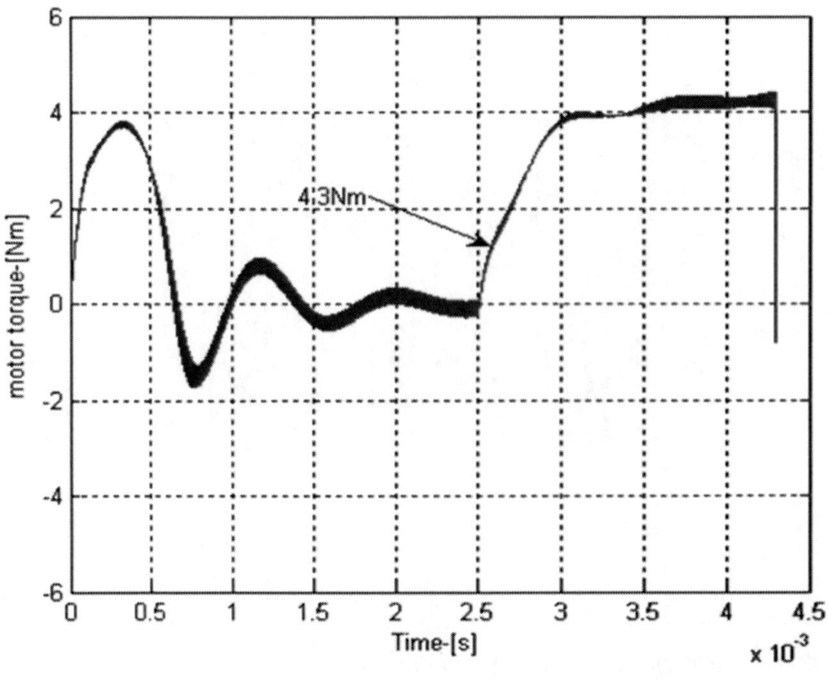

(b)

Figure 9. (Continued)

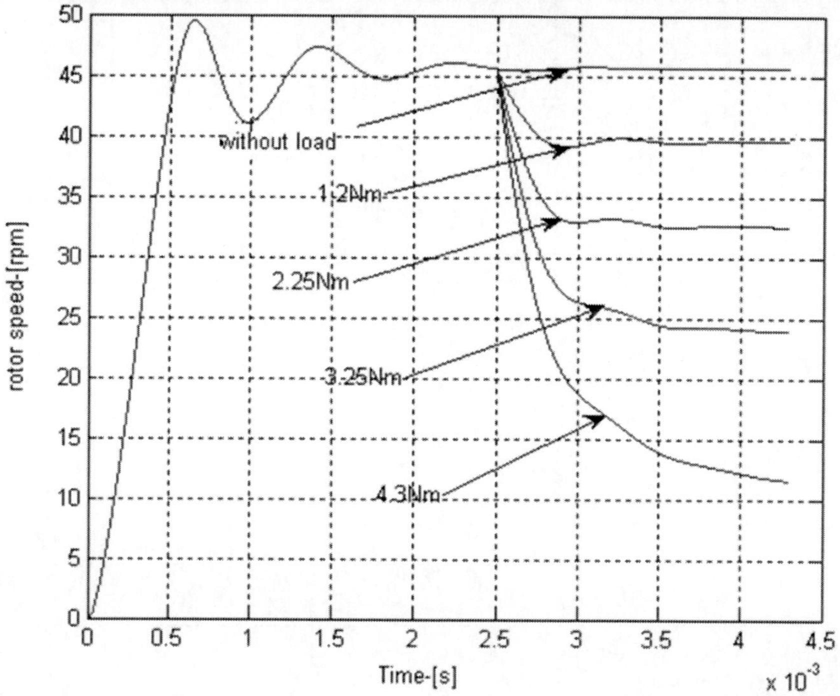

(c)

Figure 9. Motor-torque characteristics for various loads (shearing deformation neglected): (a) without load, 1.2Nm, 2.25 Nm, 3.25 Nm;(b) for a load of 4.3 Nm; and (c) rotor-speed characteristics for loads: without load, 1.2 Nm, 2.25 Nm, 3.25 Nm and 4.3 Nm.

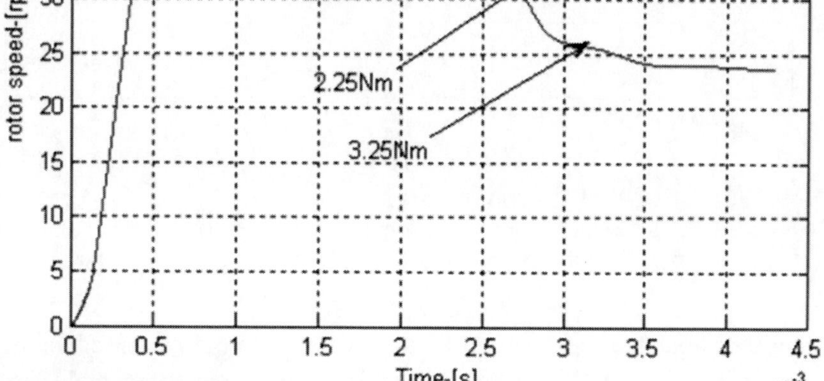

(a)

Figure 10. (Continued)

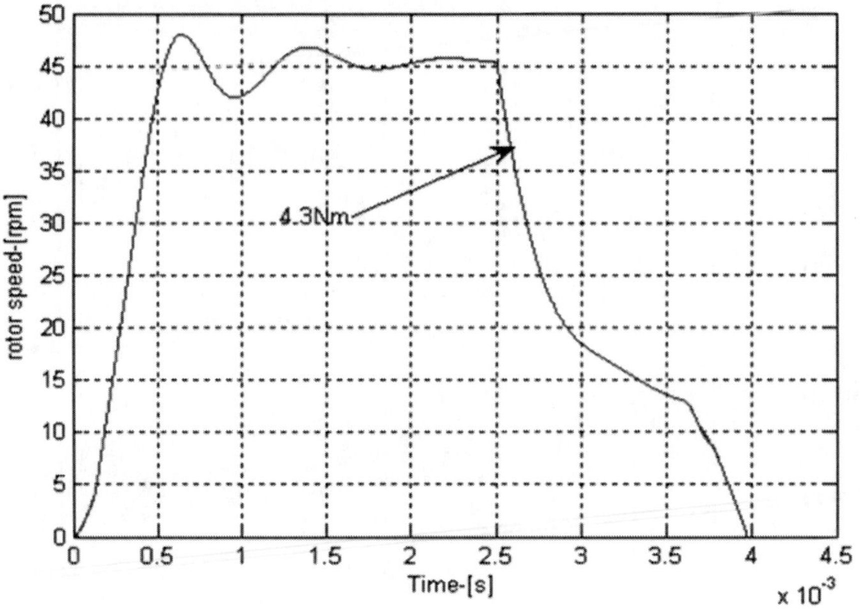

(b)

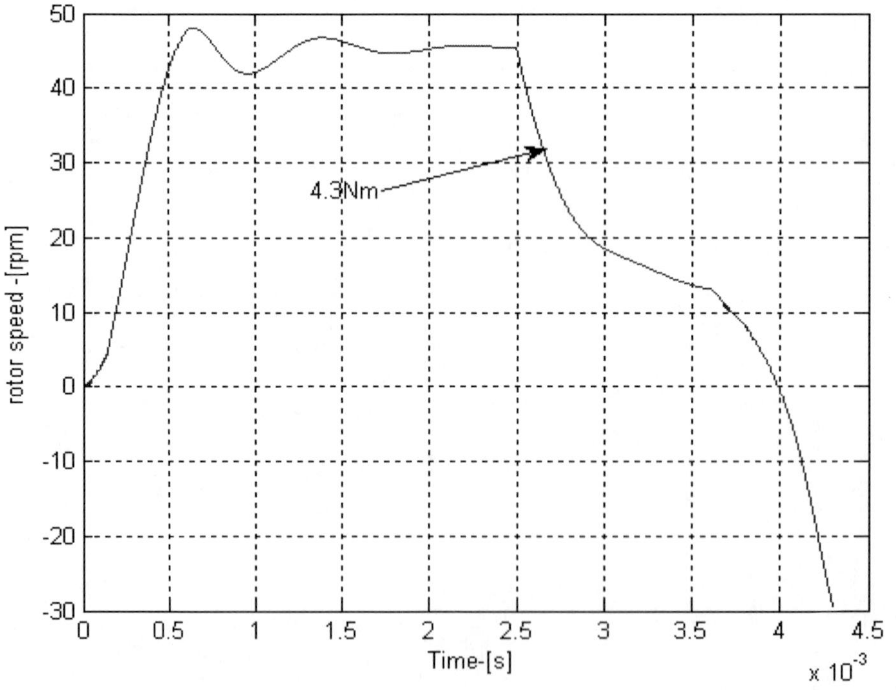

(c)

Figure 10. Rotor-speed characteristics (shearing deformation considered) for loads: (a) without load, 1.2Nm, 2.25 Nm, 3.25Nm; (b) for load: 4.3Nm; and (c) for load: 4.3Nm (with the negative speed scale).

Comparing the simulated speed-torque characteristic with the measured values allows us to verify the validity of model. Two validation rests in experimental data are used respectively, in [19] and [20].

IV.1. Test of First Validation Rest in Experimental Data

The simulation results represented by the solid line were compared with the data of the manufacturer (Figure 12). One can see that the curve followed the two measurements of the manufacturer, represented in the figure by stars and triangles [19].We distinguished the two steps in the dynamic part of the test: the starting phase and the steady state.

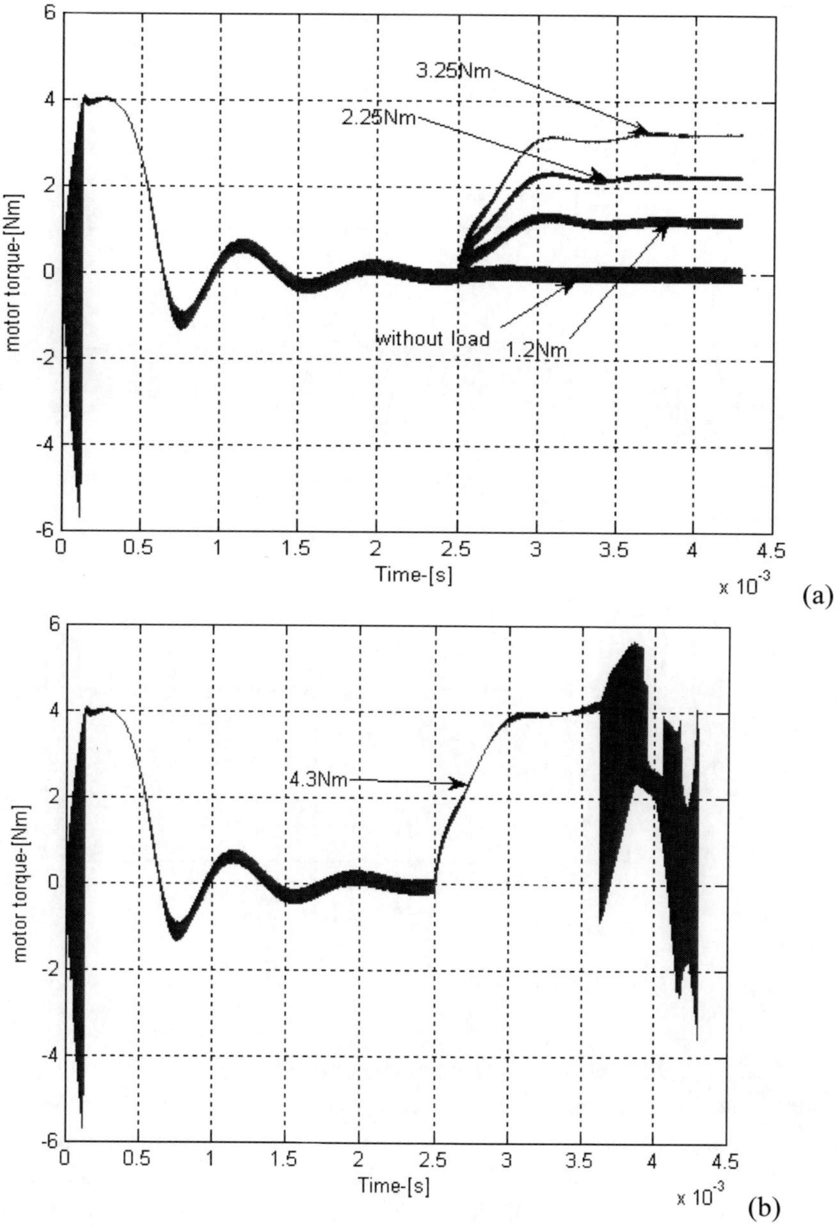

Figure 11. Motor-torque characteristics (shearing deformation considered): (a) for loads: rotor unloaded; 1.2Nm; 2.25Nm and 3.25Nm and (b) for a load of 4.3 N.

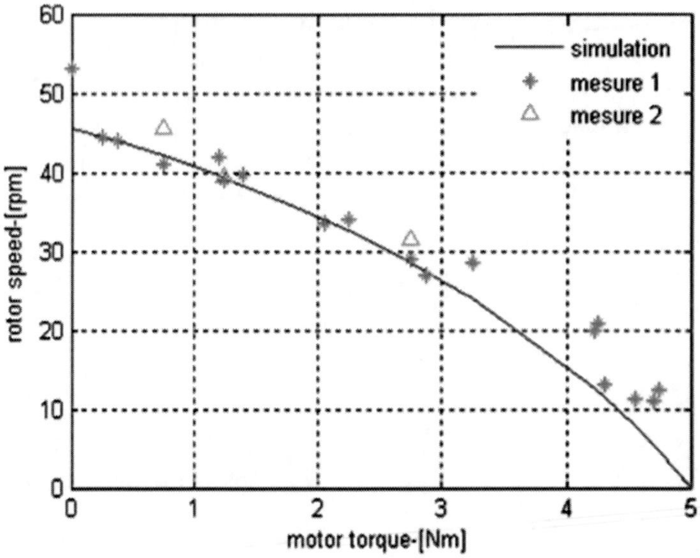

(a)

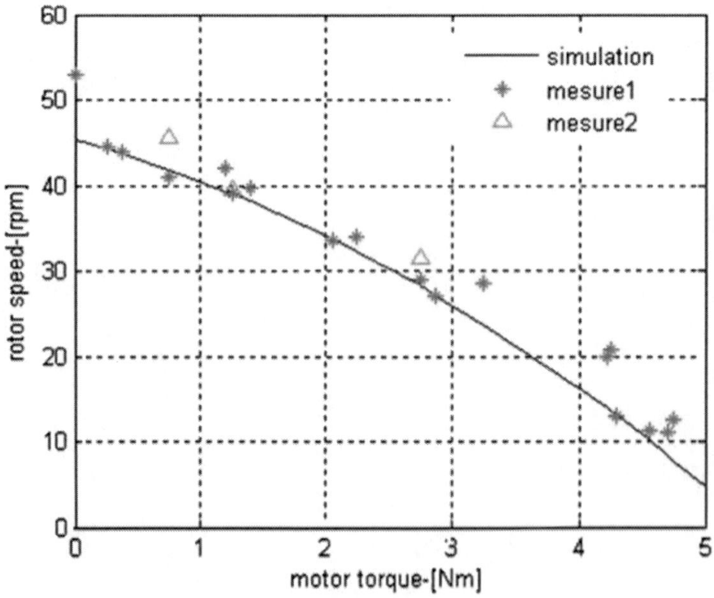

(b)

Figure 12. Measured and simulated speed-torque characteristics of the AWM90-X motor: (a) Shearing deformation neglected, (b) Shearing deformation considered.

(i) In the starting phase at the frequency used and without a load, the simulation curve is far from the measured values because of the piezo-ceramic resonance frequency (43,365 kHz see Table 2). This later takes the simulated curve too far from experimental measurements.

The optimal working frequency of the rotor for the case without a load torque is plotted. The frequency closer to that of a real working motor was equal to 46.65 kHz. It is very close to the anti-resonance frequency of the simulated mechanical system reported in [10]. For this frequency, the input current and therefore the losses in the piezoelectric ceramics are minimal. The frequency difference is attributed to the tribological phenomena acting in the zone of contact between stator and rotor. The differences between our measurements and the manufacturer's measurements, made successively, are a good example of this phenomena and the dynamic behavior of motor. Four intrinsic parameters, frequency, temperature, shearing deformation and surface state of the piezo-ceramic, originate from shifting between the two measurement series and also from the disparity between the no load measurement and the measurement taken on a loaded motor. To discover the qualitative relationship between a motor's parameters and its performance, these parameters must be taken into account in the model. The temperature variations are mainly generated by friction in the stator–rotor contact and by electrical loss in the stator. The shearing deformation taken into account in this work already results ina good correspondence between model prediction and real motor behavior. The predictions will be better if temperature and loss effects are considered.

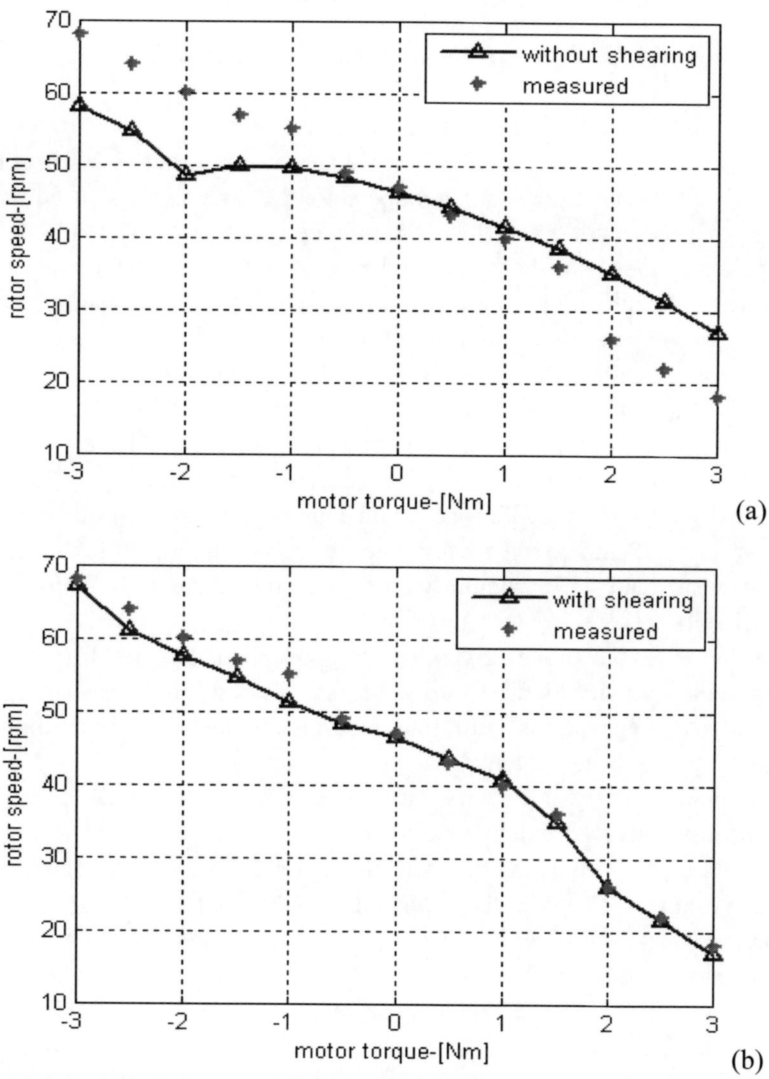

Figure 13. Measured and simulated speed-torque characteristics of the AWM90 motor; a) Shearing deformation neglected, (b) Shearing deformation considered * measured [20]; Δ simulated.

(ii) In the settling state, from 0.25Nm to 4.3 Nm, the implemented model does a good job at reflecting the real behavior of the motor. However, for load torques greater than 4.3Nm (Figurc 12(a), the shearing deformation effect and the slip phenomenon of the rotor in the interface stator/rotor are greater. The simulation values greatly differ from measured values. That is a result of the slip phenomenon effect of rotor.

When the shearing deformation (Figure 12(b)) is introduced in the model, we can see that the simulation curve approaches the first measured values particularly for loads greater than 4 Nm. The simulation curve overlaps with nine measured values. Finally, the shearing deformation effect improves the correspondence between prediction and real motor behavior. The results of the simulation, compared with the measurements made by Kandare and Wallaschek [19], are valid.

IV.2. Test of Second Validation Rest in Experimental Data

In this section, the experimental measurements made by other researchers [20] on traveling wave piezoelectric motor AWM90 type are used in order to confirm the validity of the simulation results. The simulation parameters are the same as those used in the preceding section, except that the excitation voltages amplitude was 595 V. The simulation results for the rotor speed for various loads before and after taking into account shearing deformation are shown in Figure 13. Schulte and Frhleke's measurements [20] are represented in the figure by stars, and simulation results corresponding to model with and without shearing deformation are represented by triangles connected by solid lines. The work carried out in [20] was investigated only the torque range located between−3Nmand 3Nm, because it was to be used in a speed control and the optimization of the performance of the drive system because generally, in this torque range, the analytical values of the precedent model are close to measured values. This is the case that we see in Figure 12(a) for a torque between 0Nm and 3Nm. However, most of simulated values using the preceding model are far from Schulte and Frhleke's [20] measurements, particularly in the torque ranges (−1Nm to −3Nm) and (2–3Nm) (Figure 13(a)).

However, the extended model gives a very good result (Figure 13(b)). The impact of shearing deformation acts particularly upon the speeds resulting from strong torques. The simulation results curve overlaps with four measured values and superimpose with eight measured values among thirteen points.

A point-by-point analysis of the curve reveals some asymmetry when compared with zero torque and with the measured values, due to the complicated nonlinear and time dependant behavior of system. At zero load, the two simulated values and the measured value are superimposed. From zero to 3 Nm, the extended model brings all the simulated values closer to the measured values. Left of zero, only the points corresponding to −0.5Nm and −3Nm are superimposed, and the other points are very close (overlap) with measured values. These simulation results, when compared with those measured by Schulte and Frhleke [20], are valid.

CONCLUSION

The background in the field of TWUM modeling has been discussed in this chapter. A mathematical description of the contact mechanism is developed. All the forces acting in the interface rotor/stator are introduced in the analytical model. An approach to taking into account the transient response of this motor type and the impact of shear deformation on these transients is presented. The previous extended model is simulated. The comparison between the experimental and simulated results shows a good agreement. A qualitative validation of the results obtained is provided by [19] and [20].

From the point of view of impact, the shearing deformation acts upon the starting state as well as on the settling state. The oscillations in the starting phase have been reduced in amplitude as well as in breadth. The steady state of the motor is improved. The setting time and the response time of the motor are reduced. The results are promising from the point of view of stability and applications of the actuator. This can be used in many practical applications that need a fast response such as robotic actuators and medical instruments. The overloading effect has been characterized by the slip phenomenon. The beginning of the phenomenon is delimited in time (3.5 ms) and in torque value (4.3 Nm) for prototype motor AWM90-X. These numerical results can serve as sensitive parameters in the speed control loop and predict a flexible control in comparison with the previous control systems based on previous models. Finally, validation of the extended model was performed; we notice that the small difference between measurement and simulation results originates from the tribological phenomena acting in the zone of contact between stator and rotor. The model can be improved in the future by taking into account the shearing angle (θ). This depends on the dynamics and the external forces of the motor. Shering angle can be introduced in the present model, as variable dependent on forces. One expects to further reduce the small difference between the measured and the simulated results. Consequently, we can improve the control flexibility of the actuator.

REFERENCES

[1] T. Hemsel, M. Mracek, J.T wiefel, P.Vasiljev, *ultrasonics* 44(2006) e591-e596.

[2] José M. Fernandez, Mikael Krummen and Yves Perriad, *IEEE international ultrasonics, ferroelectrics, and frequency control joint 50 th anniversary conference* 2004.

[3] Antonio Iula, Massimo Pappalardo, *IEEE transactions on ultrasonics, ferroelectrics, and frequency control*, vol. 53, no. 7, july 2006.

[4] Tomonodo Senjyo, Mitsuru Nakamura, Naomitsu Urasaki, Hideomi Sekine, Toshihisa Funabashi. Meidensha Corporation, *twenty-first annual IEEE, Applied Power Electronics Conference,* 2006. APEC apos; 06.

[5] Hamed Mojallali, Rouzbeh Amini, Roozbeh Izadi- Zamanabadi, Ali A Jalali, *ISA transactions* 46(2007) 31-40.

[6] J.T.Leinvuo,S.A.Wilson,R.W.Whatmore,M.G.Cain, *Sensors and Actuators* A133 (2007) 141-151.

[7] Y. Yano, K. Iida, K. Yabugami, Y. Nakata, T. Sakabe, *The 47th IEEE International Mid west Symposium on Circuits and Systems* 2004.

[8] K. Spanner, *physic instrumente Gmbh &Co*. K G, Karisruhe, Germany, 2006.

[9] L. Petit, et al., *IEEE Trans. Ultrason. Ferroelectr. Freq. Control* 45 (2) (1998) 277–284.

[10] L. Petit, P. Gonnard, B. Grehant, *Applications of Ferroelectrics, ISAF* 2000, 2000.

[11] L. Petit, P. Gonnard, *Sensors and. Actuators A: Phys.* 120 (1) (2005) 211–224.

[12] Lionel Petit∗, Paul Gonnard, *Sensors and Actuators A* 149 (2009) 113–119.

[13] L. Huafeng, Z. Chunsheng, G. Chenglin, *Journal of Electrical Engineering* 55 (7–8) (2004) 216–220.

[14] J. Qu, T. Zhou, *Ultrasonics* 41 (2003) 561–567.

[15] J. Qu, F. Sun, C. Zhao, *Ultrasonics* 45(2006) 22–31.

[16] H. Storck, J. Wallaschek, *International Journal of Non-linear Mechanics* 38 (2003) 143–159.

[17] Y.H. Kim, S.K. Ha, *Journal of Sound and Vibration* 263 (2003) 643–663.

[18] Y.H. Kim, S.K. Ha, *IEEE Transactions on Ultrasonics, Ferroelectrics, and Frequency Control* 50 (12) (2003).

[19] G. Kandare, J. Wallaschek, *Smart Materials & Structures* 11 (2002) 565–574.

[20] T. Schulte, N. Frhleke, *IEEE/ASME International Conference on Advanced Intelligent Mechatronics,* Como, Italy, July 8–12, 2001, pp. 504–509.

[21] F.Z. Kebbab, Z. Boumous, S. Belkhiat, *Jour Electrical systems*, Vol. N°5, 2009.

[22] F.Z. Kebbab, Z. Boumous, S. Belkhiat, Jour Electrical systems, special issue N°1, 2009.

[23] M.Djaghloul, Z.Boumous, S.Belkhiat, AIP Conf. Proc. Volume 1019,pp.420-426, *1st Mediterranean Conference on Intelligent Systems and Automation (*CISA08) Annaba June 12,2008.

[24] D. Lummer, Aufbau eines Pr¨ufstandes zur Bestimmung der station¨aren Kennlinien von Wander wellenmotore *Studienarbeit* Universit¨at-GH Paderborn, 1997.

[25] Z. Boumous, S. Belkhiat and F.Z. Kebbab, *Sensors and Actuators A* 150 (2009) 243–250.

[26] Daimler-Benz AG Druckschrift TE/P67052104000791 1991.

In: Ultrasonics: Theory, Techniques and Practical Applications ISBN: 978-1-62257-685-2
Editors: Hanako Ayabito and Mitsuko Katsukawa

Chapter 7

AUTOFOCUSING ARRAY IMAGING IN ANISOTROPIC WELD STRUCTURE

Jie Zhang*
Department of Mechanical Engineering, University Walk,
University of Bristol, Bristol, UK

ABSTRACT

The quality of an ultrasonic array image depends on accurate information about its acoustic properties. Inaccuracy of acoustic properties can cause image degradation such as blurring, mislocation of reflectors and the introduction of artefacts. In this chapter, for the specific case of an inhomogeneous and anisotropic austenitic steel weld, Monte Carlo Markov Chain (MCMC) inversion is used to estimate unknown acoustic properties from array data. The approach uses active beacons that transmit ultrasound through the anisotropic weld which is captured by a receiving array. A forward model of the ultrasonic array data is then optimised with respect to the experimental data using a MCMC inversion. The result of this process is the extraction of a material property map that describes the anisotropy distribution within the weld region. These extracted material properties are then used within an imaging algorithm, the Total Focusing Method, to produce autofocused images. This MCMC inversion approach is first applied to simulated data which is used to test the convergence, robustness and accuracy of the method and its implementation. In this case the extracted weld map is used to show improved imaging of defects within the weld as compared to an image formed assuming a constant velocity.

1. INTRODUCATION

Ultrasonic arrays have been widely used for inspecting a range of structures [1]. They offer two key advantages over single element techniques. Firstly, a given array can be used to perform a multitude of different inspections: e.g. different steering angles and focal depths.

* j.zhang@bristol.ac.uk.

Secondly, from each array location an image is formed that represents the interior of the test structure geometrically and can bc used to detect, size and locate defects in an intuitive manner. The generation of high quality ultrasonic array images is reliant on accurate knowledge of the array element locations as well as the acoustic properties of the test structure. The element locations are defined by the array manufacturing process and so are not usually thought of as unknowns. The acoustic properties of the test structure, principally velocity, are either obtained from prior knowledge of similar materials, or experimentally. However, neither of these approaches is particularly effective in the context non-homogeneous and/or anisotropic materials such as austenitic steel welds. The distribution of anisotropy amplitude and orientation in a weld is often presented in the form of a weld map and typically this information is uncertain for a given weld. Currently most ultrasonic inspections of such welds assume a single velocity, based on that found in the parent metal. When a weld map is used, uncertainties in the weld map lead to errors in estimating the time-of-flight of ultrasonic waves through the weld which in turn causes image degradation and ultimately adversely affects detection and characterisation of defects (e.g. Chassignole et al. [2]) .

In this chapter, an array based beacon system is described and used in conjunction with a Monte Carlo Markov Chain (MCMC) inversion scheme to extract weld maps. The approach is to measure the time-of-flight of ultrasonic signals that propagate from active beacons, travel through the weld and are recorded by a receiving array. The errors between these measurements and a forward model of wave propagation in the weld are minimised using Bayesian estimation of the unknown model parameters via a MCMC optimisation scheme. The forward model of the wave propagation discretises the weld into a number of homogeneous sub-regions, each with variable anisotropy, and the propagation through this medium modelled using a fast ray tracing algorithm.

2. MODELS

In this section, Monte Carlo Markov Chain (MCMC) optimisation techniques are introduced and their application to weld parameter estimation described. Figure 1 shows the MCMC inversion process schematically. The process starts with a set, $\mathbf{x}'$, of proposed weld parameters which are used in a forward ray tracing model to compute time-of-flights between active beacons and a receiver array. A Bayesian Monte Carlo estimator is used to iteratively determine the optimal weld parameters by minimising the difference between $\hat{t}_{mn}$ (experiment) and t_{mn} (model), i.e., $\Delta t_{mn} = \hat{t}_{mn} - t_{mn}$, until Δt_{mn} convergences to a predefined level. Note that subscripts m and n refer to beacon and receiver elements respectively. The steps of using MCMC inversion to estimate weld parameters can be briefly described as,

1. Provide initial estimate of weld parameters, $\mathbf{x}'$
2. Simulate forward wave propagation through the weld
3. Compare simulation with actual measurements
4. Either accept or reject this estimate for inclusion in probability calculation

5. Generate new estimate based on last accepted estimate
6. Repeat steps 2 to 5 until enough data exists to make reliable estimate of the probability distribution of weld parameters given the measured data.

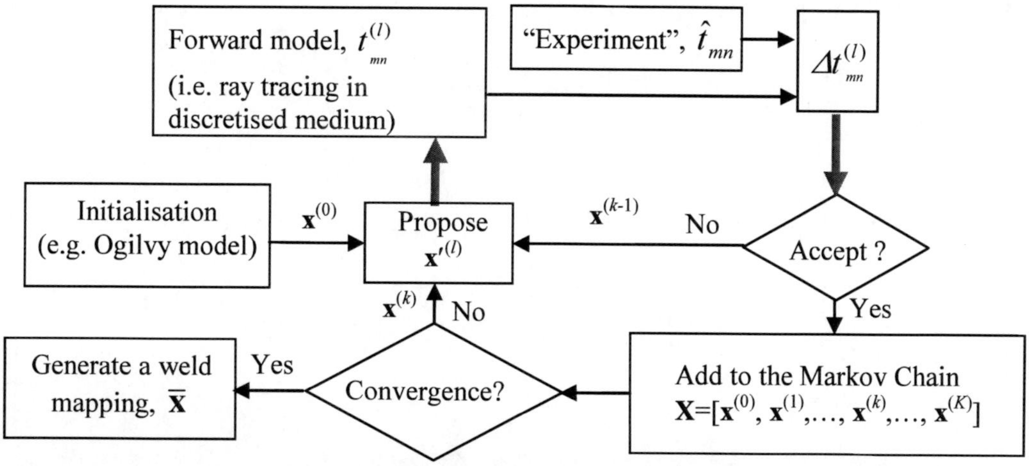

Figure 1. Diagram of the estimation of weld model parameters based on Monte Carlo Markov Chain (MCMC) inversion of time-of-flight data.

The details of each model used in the above procedure are described in the following sections.

2.1. Discretised Model of the Anisotropic Weld

The austenitic weld is discretised modelled as a grid of homogeneous but anisotropic sub-regions. In general the anisotropy in each sub-region can be characterized by a small number of parameters and it is relatively straight forward to model the wave propagation through the various sub-regions by ray tracing. Figure 2(a) shows an example of a possible weld anisotropy distribution using 12×18 anisotropic sub-regions, where each sub-regions is 5×5 mm^2 and the total region size is 60×90 mm^2. The anisotropy in each sub-region is characterised by an orientation angle which gives the principal direction and an amplitude which indicates the anisotropy level as a fraction of the nominal speed. The velocity distribution, $c_{ij}(\theta)$, in each sub-region is approximated by a sinusoidal variation,

$$c_{ij}(\theta) = c_0\left[1 + A_{ij}\cos\left(2\left(\theta - \phi_{ij}\right)\right)\right], \tag{1}$$

where i and j refer to the x and z locations of the sub-region, A_{ij} is the amplitude of the anisotropy as a fraction of the nominal speed c_0, ϕ_{ij} is the angle corresponding to the direction of maximum velocity (grain orientation angle) and θ and ϕ_{ij} are defined counter-clockwise relative to the positive z-axis. Figure 2(a) shows an example velocity distribution with angle θ, when c_0 = 6000 m/s, A = 0.2 and ϕ = 45°. It should be noted that, although this simplified two-parameter (A and ϕ) anisotropy model has no physical basis, it is a reasonable fit to reality. However, the two-parameter anisotropy weld model can be replaced by any weld model derived from the physics [2] in the Monte Carlo inversion process if required.

Various models of the distribution of anisotropy in austenitic weld are available [2-7]. In this chapter we select the Ogilvy model for a V-shaped weld which is one of the simplest, and in which the grain orientations $\phi_{ij}\ (x, z)$ are modelled by [1],

$$\phi_{ij}(x,z) = \tan^{-1}\left(\frac{W_1/2 + (z - Z)\tan\alpha}{x}\right) + \frac{\pi}{2}, \tag{2}$$

where Z is the weld depth, W_1 is the width of the weld root, and α is the grain angle at the fusion face. The anisotropy amplitude, $A_{ij}(x, z)$ is a non-zero constant A_0 within the weld and zero elsewhere,

$$A_{ij}(x,z) = \begin{cases} A_0 & -(z\text{-}C)/D \le x \le (z\text{-}C)/D,\ -Z \le z \le 0 \\ 0 & \text{elsewhere} \end{cases} \tag{3}$$

where $D = 2Z/(W_2\text{-}W_1)$ is the gradient and $C = -ZW_2/(W_2\text{-}W_1)$ is the z-axis intersect of the fusion faces, and W_2 is the width of the weld cap. Figure 2(b) shows a typical weld map, in this case the result of using the Ogilvy equations 2-3 and $c_0 = 5800$, $A_0 = 0.1$, $Z = 60$ mm, $W_1 = 5$ mm, $W_2 = 40$ mm, and $\alpha = 23°$. Based on this weld model, the weld parameters **x'** used in the Monte Carlo inversion process are given by **x'** $= [A_{11}, A_{12}, \cdots, A_{ij}, \phi_{11}, \phi_{12}, \cdots, \phi_{ij}]$.

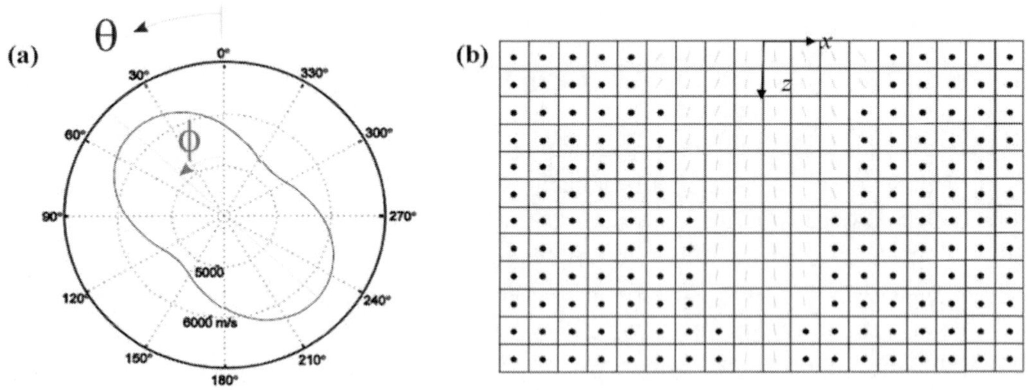

Figure 2. Discretised model of the anisotropic weld: (a) an example group velocity profile for the two-parameter anisotropy model and (b) anisotropy strength and orientation parameters for the simulated weld based on the Ogilvy model; the length and orientation of the lines represent the strength and orientation of the anisotropy.

2.2. Time-of-flight Estimation

The times-of-flight are measured from the received array data $e_{mn}(t)$ by cross-correlating each of the time-traces with the known transmitted signal and locating the maximum correlation values. The time-of-flight measurement is given by,

$$\hat{t}_{mn} = \arg\max\{R_{mn}(\tau)\}\ , \tag{4}$$

where,

$$R_{mn}(\tau) = \frac{\int_{-\infty}^{\infty} e_{mn}(t) s(t-\tau) dt}{\sqrt{\int_{-\infty}^{\infty} e_{mn}^2(t) dt} \sqrt{\int_{-\infty}^{\infty} s^2(t) dt}}\ , \tag{5}$$

is the normalised cross-correlation function and $s(t)$ is the known transmitted signal. The confidence in the measurement $\hat{t}_{mn}$ can be quantified by the correlation coefficient,

$$\rho_{mn} = R_{mn}(\hat{t}_{mn})\ , \tag{6}$$

which varies between a value of zero for no correlation and one for perfect correlation.

Using correlation in this way allows the time-of-flight to be accurately estimated from the simulated array data as there is only a single signal present and the noise is negligible. However, in experiments, the array data is distorted due to transmission through the anisotropic weld, noise due to the effects of grain scattering and multi-path arrivals [3, 4]. Note that the multi-path effects are particularly problematic for the correlation based time-of-flight technique as they result in interference which distorts the shape of the received signals. For this reason the experimental received signals are processed by the CLEAN algorithm [5-6] that extracts the signal corresponding to the first arrival.

The CLEAN algorithm operates by making the assumption that the multi-path signal is composed of a finite collection of scaled, delayed, and phase-shifted replicas of the transmitted signal [5-6],

$$E(f) = S(f)\left(\sum_{n=1}^{N} A(n)\exp(j(-2\pi f t(n) + \phi(n)))\right)\ , \tag{7}$$

where $E(f)$ is the multi-path signal spectrum, $S(f)$ is the transmitted signal spectrum, and $A(n)$, $t(n)$, and $\phi(n)$ are the amplitude, time delay, and phase of the nth replica component. The initial residue is set equal to the total received signal,

$$R_0(f) = E(f)\ . \tag{8}$$

In each iteration the algorithm locates the maximum amplitude of the residual signal envelope,

$$A(n) = \max\{|\tilde{r}_{n-1}(t)|\} \tag{9}$$

where,

$$r_n(t) = \int_{-\infty}^{\infty} R_n(f) \exp(j2\pi ft) df \tag{10}$$

and the tilde denotes the Hilbert transform, and the corresponding time delay and phase values at the maximum are measured as,

$$t(n) = \arg\min \{ |\tilde{r}_{n-1}(t)| - A_n \} \tag{11}$$

$$\phi(n) = \text{angle}\{ r_{n-1}(t_1) \} \tag{12}$$

The nth multi-path component can then be defined by,

$$S_n(f) = A(n) \exp(j(-2\pi ft(n) + \phi(n))) S(f) \,, \tag{13}$$

and the residue is updated by subtracting this component,

$$R_n(f) = R_{n-1}(f) - S_n(f) \,. \tag{14}$$

Although it is helpful to know *a priori* the number of delayed replicas N in the multi-path signal, it is not strictly necessary. It is possible to define a minimum amplitude threshold and to terminate the algorithm when the amplitude of the nth component $A(n)$ drops below this threshold.

2.3. Monte Carlo Inversion of Time-of-flight Data

2.3.1. Dijkstra Algorithm for Fast Ray Tracing

Ultrasonic wave propagation can be modelled efficiently in inhomogeneous anisotropic media by ray tracing. The shooting method has been used elsewhere for austenitic weld modelling [3, 4], whereby trial rays are propagated from the source and the initial angle is optimised until the ray intersects the destination point. The bending method has also been used for austenitic weld modelling, whereby a parameterised ray path (i.e., a spline curve) is defined between the source and destination point and the path parameters are optimised to achieve the minimum travel time according to Fermat's theorem. However, both of these methods are considered to be too slow in the context of the MCMC approach, where thousands or millions of ray-tracing operations must be performed.

The Dijkstra algorithm was originally proposed as an efficient means of solving the "travelling salesman problem" in graph theory. In the travelling salesman problem, given a list of cities and their locations, the task is to find a shortest possible tour that visits each city exactly once. Solutions to this problem have since been applied to a number of analogous problems, including efficient ray tracing in seismology [7]. The Dijkstra algorithm for ray tracing models the space as a "graph", i.e., a collection of nodes, and propagation between nodes is permitted only via connections between sets of neighbouring nodes. This is illustrated in Figure 3 for a rectangular grid of nodes and permitting connection to 3-levels of neighbouring nodes. The algorithm operates by systematically testing all possible travel times to each node from a specified source node to obtain the global minimum travel times. In this way, the algorithm finds the ray paths (i.e. routes that take the minimum time) from the source to every point in space (including the receiver positions) and the algorithm is repeated for each beacon. It can be seen that the Dijkstra algorithm provides an approximate quantized ray solution. However, the approximation error can be reduced to an arbitrary level at the cost of increased computation time by employing a finer grid of nodes and a larger number of permitted neighbour connections.

2.3.2. Bayesian Estimation via Monte Carlo Markov Chains

The discretised medium model (section 2.1) can be used as a forward model to approximate the times-of-flight from the active beacons, through the anisotropic weld to the receiver array. The inverse problem is to estimate the unknown model parameters, **x,** from the measured time-of-flight data, $\hat{\mathbf{t}} = \left[\hat{t}_{11}, \hat{t}_{12}, \cdots, \hat{t}_{mn}\right]$. A probabilistic solution to the inverse problem can be obtained using Bayes' theorem [9],

$$\mathbf{P}(\mathbf{x} \mid \hat{\mathbf{t}}) = \frac{\mathbf{P}(\hat{\mathbf{t}} \mid \mathbf{x})\mathbf{P}(\mathbf{x})}{\mathbf{P}(\hat{\mathbf{t}})} \tag{15}$$

where $\mathbf{P}(\mathbf{x} \mid \hat{\mathbf{t}})$ is the posterior distribution, $\mathbf{P}(\hat{\mathbf{t}} \mid \mathbf{x})$ is the likelihood function, $\mathbf{P}(\mathbf{x})$ is the prior distribution, and $\mathbf{P}(\hat{\mathbf{t}})$ is a normalizing constant. The posterior distribution is the conditional probability of the model parameter values given the measurements and the assumed prior knowledge; once obtained, it can be analysed to extract model parameter estimates, e.g., using the most probable solution (maximum *a priori* or maximum likelihood estimation) or using statistical moments, such as the mean. The likelihood function expresses the conditional probability of the measurements with respect to the model parameters and here we assume independent Gaussian-distributed errors on the time-of-flight measurements,

$$\mathbf{P}(\hat{\mathbf{t}} \mid \mathbf{x}) = \prod_{m,n} \mathbf{P}(\hat{t}_{mn} \mid \mathbf{x}), \tag{16}$$

where

$$\mathbf{P}(\hat{t}_{mn} \mid \mathbf{x}) = \exp\left(-\frac{\left(\hat{t}_{mn} - t_{mn}\right)^2}{\sigma^2}\right), \tag{17}$$

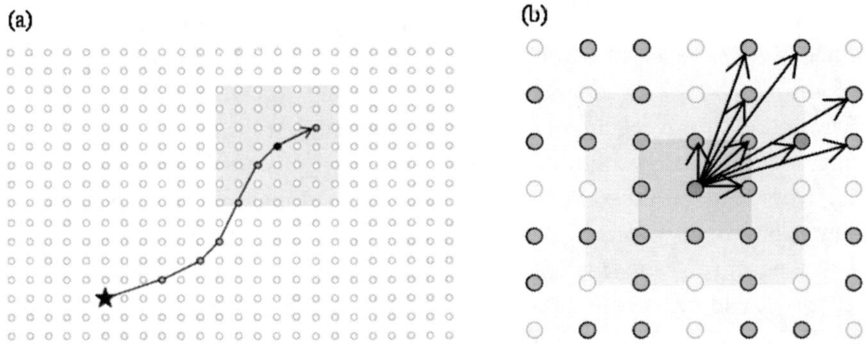

Figure 3. Ray tracing geometry for the Dijkstra algorithm using (a) a rectangular grid of nodes and (b) 3-level neighbour connectivity.

is the likelihood of a single measurement, $\hat{t}_{mn}$, with respect to the actual travel time t_{mn} and σ^2 is the variance. Note that it is assumed that, for the prior distribution, the model parameters have equal probability over a range of values $[0, A_{\max}]$ and $[\phi_{\min}, \phi_{\max}]$, which leads to,

$$\mathbf{P}(\mathrm{x}) = \prod_{i,j} \mathrm{P}(A_{ij})\mathrm{P}(\phi_{ij}) \tag{18}$$

where

$$\mathrm{P}(A_{ij}) = \begin{cases} 1/A_{\max} & 0 \le A \le A_{\max} \\ 0 & \text{elsewhere} \end{cases}, \quad \mathrm{P}(\phi_{ij}) = \begin{cases} 1/(\phi_{\max} - \phi_{\min}) & \phi_{\min} \le \phi \le \phi_{\max} \\ 0 & \text{elsewhere} \end{cases} \tag{19}$$

The posterior distribution is difficult to compute due to the multi-dimensional integration of the likelihood function $\mathbf{P}(\hat{\mathbf{t}} \mid \mathbf{x})$ over the model parameters, **x**, required to obtain the normalizing constant $\mathbf{P}(\hat{\mathbf{t}})$. The Metropolis-Hastings algorithm is a MCMC technique that can generate samples from any probability distribution given that a function proportional to the distribution can be computed. Thus, rather than computing the posterior distribution directly, the Metropolis-Hastings algorithm can be used to generate a set of solutions for the model parameters (from the known likelihood function and prior distribution) that are distributed according to the posterior distribution; this set of solutions can be used to approximate the posterior distribution and to estimate the model parameters [3].

The Metropolis-Hastings algorithm operates by generating a Markov chain; a discrete random process in which each state is dependent upon the previous state only. The Markov chain is a sequence of possible solutions for the model parameters $\mathbf{X}=[\mathbf{x}^{(0)}, \mathbf{x}^{(1)},\ldots, \mathbf{x}^{(k)},\ldots, \mathbf{x}^{(K)}]$. At each iteration, a candidate solution $\mathbf{x}'$ is generated from a proposal distribution $\mathbf{P}(\mathbf{x}' \mid \mathbf{x}^{(k-1)})$ with dependence on the previous solution in the Markov chain, $\mathbf{x}^{(k-1)}$. The candidate solution is then either accepted or rejected according to the acceptance probability distribution ,

$$\mathbf{P}\left(\mathbf{x}^{(k)}=\mathbf{x}'\,|\,\mathbf{x}^{(k-1)}\right)=\min\left(\frac{\mathbf{P}\left(\hat{\mathbf{t}}\,|\,\mathbf{x}'\right)\mathbf{P}\left(\mathbf{x}'\right)}{\mathbf{P}\left(\hat{\mathbf{t}}\,|\,\mathbf{x}^{(k-1)}\right)\mathbf{P}\left(\mathbf{x}^{(k-1)}\right)},1\right). \tag{20}$$

It can be proved that the distribution of solutions in the resulting Markov chain $\mathbf{X}$ converges to the posterior distribution $\mathbf{P}\left(\mathbf{x}\,|\,\hat{\mathbf{t}}\right)$, where convergence is attained once the chain becomes stationary, i.e., when the distribution of solutions is independent of the sequence index k. The algorithm can be started using an arbitrary initial solution $\mathbf{x}^{(0)}$. However, convergence will be quicker if an initial solution is chosen that is close to the true solution.

For the discretised weld model, a suitable proposed distribution selects a sub-region ij at random with uniform probability and alters its anisotropy parameters with uniform probability over a range of values $[-\Delta A/2, \Delta A/2]+A_{ij}^{(k-1)}$ and $[-\Delta\phi/2, \Delta\phi/2]+\phi_{ij}^{(k-1)}$. The Metropolis-Hastings algorithm for Markov chains is summarised below,

1. Let previously accepted solution be $\mathbf{x}^{(k-1)}$
2. Propose new candidate solution $\mathbf{x}'$ for weld parameters by randomly altering one or more parameters in $\mathbf{x}^{(k-1)}$
3. Use the forward model to simulate arrival times $\hat{\mathbf{t}}(\mathbf{x}')$
4. Estimate likelihood, $\mathbf{P}\left(\hat{\mathbf{t}}\,|\,\mathbf{x}'\right)$, of obtaining actual measured arrival times, $\hat{\mathbf{t}}(\mathbf{x}')$, if $\hat{\mathbf{t}}(\mathbf{x}')$ were the true arrival times using the assumed measurement error distribution
5. Is likelihood of proposed solution, $\mathbf{P}\left(\hat{\mathbf{t}}\,|\,\mathbf{x}^{(k)}\right)$, more than likelihood of previous solution, $\mathbf{P}\left(\hat{\mathbf{t}}\,|\,\mathbf{x}^{(k-1)}\right)$?
   - Yes: accept it and add to chain ($\mathbf{x}^{(k)}=\mathbf{x}'$)
   - No: accept it with probability $\dfrac{\mathbf{P}\left(\hat{\mathbf{t}}\,|\,\mathbf{x}'\right)}{\mathbf{P}\left(\hat{\mathbf{t}}\,|\,\mathbf{x}^{(k-1)}\right)}$ or reject it.

3. Monte Carlo Markov Chain Inversion of Simulated Data

In this section, the performance of the MCMC inversion scheme is explored using simulated data based on the array and test structure configuration shown in Figure 4. Eight active point-like beacons are positioned above the weld at intervals of 15 mm and the 60 element receivers at intervals of 2 mm are positioned below the weld. The weld and surrounding parent material was discretised in to a grid of 12x24 square elements, each 5x5 mm^2 in size and each represented by the two parameter model of anisotropy. For this simulation the Ogilvy weld map (equations 1-3) was assumed with $c_0 = 5800$, $A_0 = 0.1$, $Z = 60$ mm, $W_1 = 5$ mm, $W_2 = 40$ mm, and $\alpha = 23°$. However, it should be noted that other weld maps could easily be used and the emphasis is on implementing a general approach. The time-of-flight from all possible combinations of beacon and receiver element pairs were computed using the Dijkstra fast ray tracing algorithm (section 2.3.1) with a 2 mm square grid and allowing paths that include the 3 nearest neighbours. Random timing errors caused by Gaussian white noise were added to the extracted time-of-flight data to better simulate experiment. As well as the effect of noise, the influences of the sub-region size and the initial conditions (i.e. the starting point of the iteration process) were also investigated.

In the MCMC inversion scheme the known variance of the measurement noise was used in the likelihood function and a broad prior distribution was assumed: $\phi_{min} = -90°$, $\phi_{max} = 90°$ and $A_{max} = 0.2$. For the proposal distribution, local search ranges of ±10% of the maximum ranges were defined, i.e., $\Delta A = 0.2 \times A_{max}$ and $\Delta\phi = 0.2 \times (\phi_{max} - \phi_{min})$. The algorithm was started with a crude initial estimate, termed 'case 1', which is shown in Figure 5(a) and given by,

$$\phi_{ij}^{(0)}(x,z) = \begin{cases} -30° & x \le 0 \\ 30° & x > 0 \end{cases} \tag{21}$$

and

$$A_{ij}^{(0)}(x,z) = \begin{cases} 0.1 & |x| \le 20\,\text{mm} \\ 0 & \text{elsewhere} \end{cases} \tag{22}$$

The MCMC scheme was run for 150,000 proposal solutions to capture enough acceptable solutions for further analysis; each Dijkstra ray-tracing step (i.e. one beacon to all receivers) took 2 seconds and in total each inversion took 80 hours to run on a standard desktop computer.

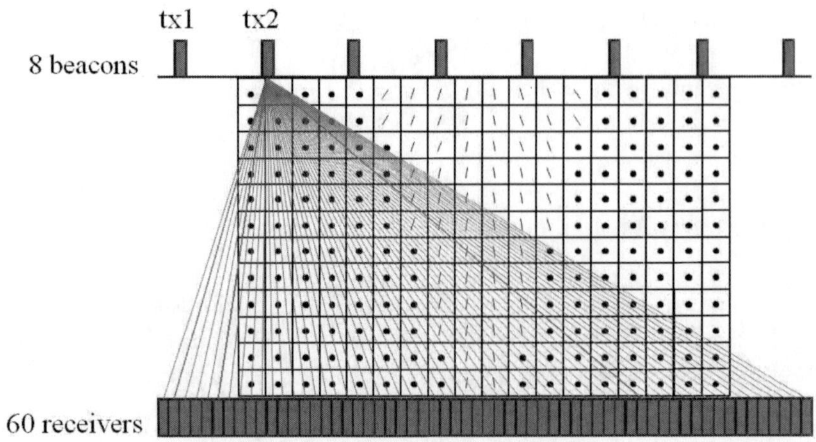

Figure 4. Beacon and receiver array configuration used to verify the Monte Carlo Markov Chain (MCMC) inversion approach. Note that the weld is shown with a weld map obtained using the Ogilvy model.

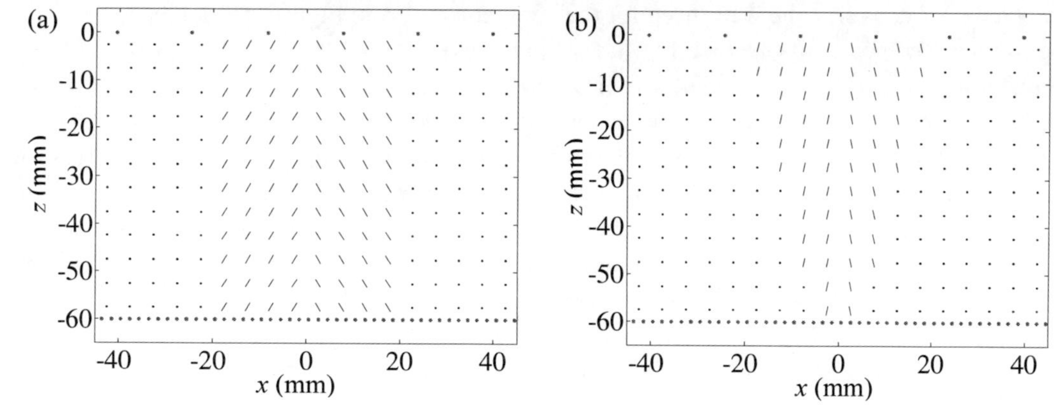

Figure 5. Crude initial condition (case 1).

The evolution of the RMS error on the time-of-flight estimates (Δt_{mn}) is shown in Figure 6(a). The error decreases from a value of approximate 80 ns for the initial estimate down to a steady-state value of approximately 20 ns. This error is equal to the known simulated measurement error. The resulting Markov chain consisted of approximately 25,000 accepted solutions. The first 10,000 accepted solutions were in the pre-convergence stage and so were discarded. The remaining independent 15,000 solutions were in the post-convergence stage (i.e. after 40,000 iterations) and were used as representative samples from the posterior distribution. The mean values of the posterior parameter distributions were used as the best estimates of the true values and the resulting weld map estimate is shown in Figure 6(b).

The samples from the posterior distribution were analysed to obtain estimates of the weld parameters. The sample distributions of A and ϕ within 3 different model sub-regions, marked as locations A-C in Figure 6(b), are shown in Figure 6(c) and (d). From Figures 6(c) and (d), it can be seen that there is generally reasonable agreement between extracted and true values for both A and ϕ. The extracted parameters at location A exhibit the poorest agreement with the true values and the greatest statistical spread. This is thought to be because the locations B and C are illuminated by greater number of rays from a greater number of incident directions when compared to location A, which leads to more reliable inversion.

The robustness of the MCMC approach is demonstrated by the convergence of mean error and variance of the measured weld maps for various initial conditions and added noise levels. Table 1 shows the four initial conditions used in this study. The initial condition cases 2-4 all assume knowledge of the location of the weld region and so limit the search space to the known welded region. Case 2 (like case 1) is also a crude estimate of the weld map but now includes assumed knowledge of the welded area. Case 3 uses the Ogilvy weld map so can be thought of as the perfect initial condition. Case 4 also uses the Ogilvy weld map, but is based on a finer, 1.5x1.5 mm^2 grid of sub-regions. In addition to varying the initial conditions, a number of different levels of noise were added in combination with initial condition case 1.

Figure 7 shows a comparison of weld map errors, which are quantified by the mean and variance of the difference between the weld maps extracted using MCMC inversion and the known Ogilvy weld map (shown in Figure 4). Note that the statistics were taken from the central weld region only, marked as the dashed box in Figure 6(b), and only solutions accepted post-convergence were used for the comparison. From Figure 7(a) it can be seen that in all cases the mean anisotropy amplitude is predicted within 22% of the true value (0.1). It can also be seen that the combined use of the perfect initial condition and a finer grid reduce the mean errors to 2%. It is also apparent that the variance is quite high (in the range 22-37%) and this is thought to reflect the varying levels of confidence in the extracted values across the central region due to spatial distribution of ray paths. Further work is required to quantify this spatial variation and devise optimal arrangements of beacons and receivers. Figure 7(b) shows that the mean error in the extracted anisotropy orientation is less than 2° in all cases, again with a relatively high variance of between 3-9°.

Table 1. Initial condition cases used in the model to estimate weld maps

Label	1	2	3	4
Weld region only	No	Yes	Yes	Yes
Model (Eqs)	21-22	21-22	2-3	2-3
Sub-region size (mm)	5×5	5×5	5×5	1.5×1.5

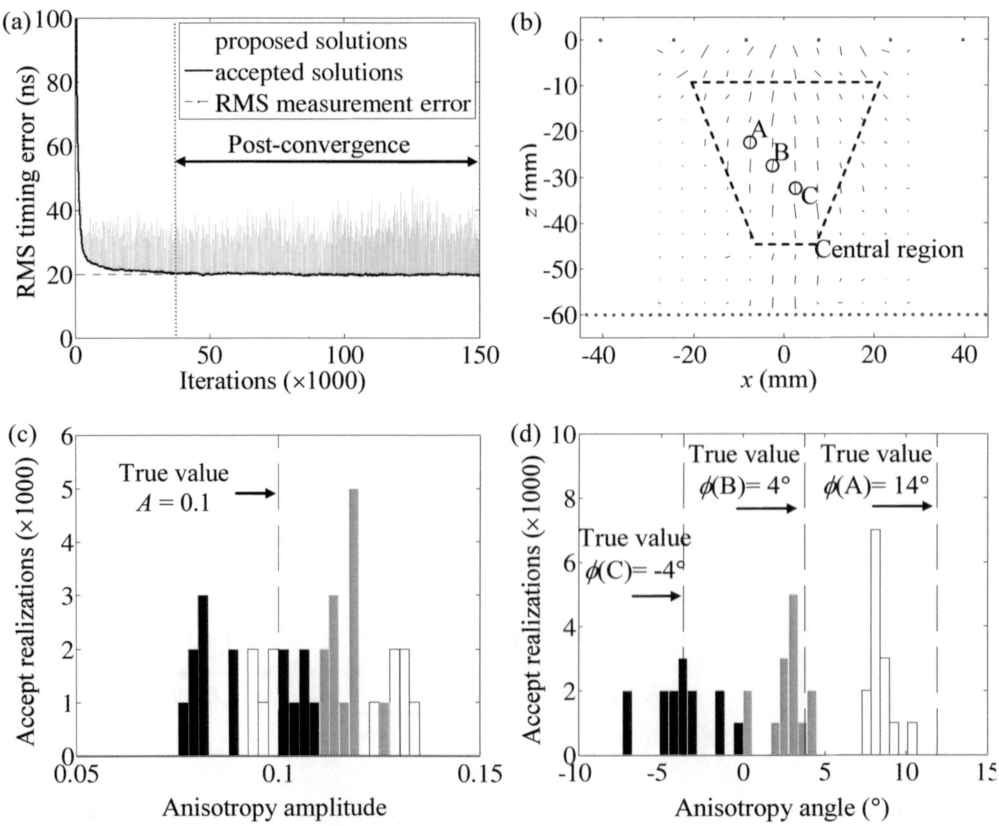

Figure 6. Estimated MCMC results from the simulated data using the case 1 initial condition and 20 ns noise level, (a) evolution of the RMS travel time error, (b) estimated mean weld map, and (c, d) estimated posterior distributions of anisotropy strength and orientation parameters in the regions shown in (b) – A(white), B(grey) and C(black).

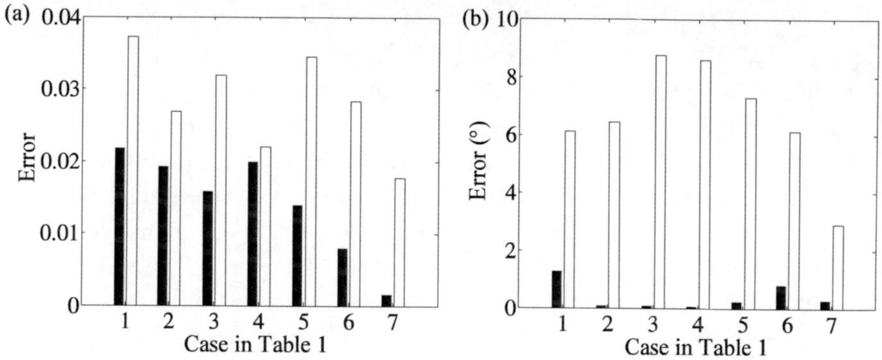

Figure 7. The effect of initial condition and added noise level (see Tables X and Y) on estimated weld maps; (a) anisotropy strength and (b) anisotropy orientation. Mean and RMS errors are indicated as the black and white bars in each figure. Note the numbers on the refer tot eh initial condition case (see Table X) and the letters the added noise level; A=10 ns, B=20 ns, c=50 ns and D= 100 ns.

Table 2. Array parameters

Array parameter	Value
Number of elements	2 x 64
Element width (mm)	1.25
Element pitch (mm)	1.5
Element length (mm)	15
Centre frequency (MHz)	2
Bandwidth (-6 dB) (MHz)	1-3

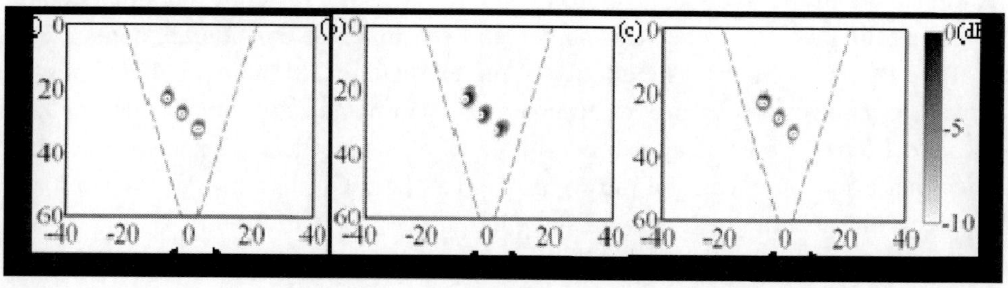

Figure 8. Simulated Total Focusing Method images using an array placed to the side of the array (see Table X for array specification); (a) assuming constant velocity, c=5800 m/s, and (c) using the extracted weld map shown in Figure 6(b).

Table 3. Locations and its errors of point scatterers A, B and C in Figure 8

Point scatterer	Location in Figure						Error in Figure 8(b)	
	8(a)		8(a)		8(b)			
	x	*z*	*x*	*z*	*x*	*z*	*x*	*z*
A	-7.5	22.5	-7	23.5	-7.5	22.5	0.5	1
B	-2.5	27.5	-1.5	28	-2.5	27.5	1	0.5
C	2.5	32.5	4	32.5	2.5	32.5	1.5	0

As a final demonstration the extracted weld map was then used to simulate an inspection in which an array (2MHz, 64 elements spaced at $\lambda/2$, see details in Table 2) was placed to the left of the weld and a number of point reflectors located within the weld imaged using the Total Focusing Method [8]. The estimated mean weld map by using the case 4 initial condition and 20 ns noise level (as shown in Figure 8(a)) was used to correct image, and the imaging results are shown in Figures 8(b) and (c). In Figure 8(b) the reflectors are imaged assuming that the weld is isotropic and has the same velocity as the parent metal. From Table 3 it can be seen that this results in location errors of up to 1.5 mm. It can also be seen from Figure 8(c) and Table 3 that the use of the extracted weld map (as shown in Figure 8(a)) results in the correct location of the reflections to within 0.1 mm. Taking the results of Figures 6-8 together, the authors believe this demonstrates pragmatically that the inversion process is stable and the quality of the extracted weld map is good enough to significantly improve imaging performance.

CONCLUSION

An approach based on a Monte Carlo Markov Chain (MCMC) inversion scheme for anisotropic material property extraction has been proposed and its application to improve images of austenitic welds has been investigated. A series of active beacons are used to propagate ultrasound through the weld and this is then recorded at a receiver array. A forward model that combines a discretised model of the weld with a fast ray tracing algorithm is then iteratively compared to the experimental data within the MCMC inversion scheme to estimate the weld map.

The robustness of the MCMC inversion approach was first tested using simulated data. It was shown that the inversion process is stable, and for the cases considered, converges in the vicinity of the correct solution. The quality of the extracted weld map was then shown to be good enough to significantly improve the imaging performance. In this simulation, locations errors of up to 1.5 mm were reduced to negligible levels using the extracted weld map. It was also shown that prior knowledge of the weld map can be incorporated into the approach and used to increase the accuracy of the extracted weld map.

REFERENCES

[1] B. W. Drinkwater and P. D. Wilcox, *NDE E Int.,* 2006, vol. 39, pp. 525-541.

[2] J. Ogilvy, *Ultrasonics,* 1986, vol. 24, pp. 337-347.

[3] M. Sambridge and K. Mosegaard, *Rev. Geophys.,* 2002, vol. 40(3), pp. 1-29.

[4] G. Connolly, M. Lowe, J. Temple and S. Rokhlin, *J. Acoust. Soc. Am.,* 2010, vol. 127, 2802-2812.

[5] P. Gough, IEEE Trans. *Signal Process.,* 1994, vol. 42, 1312-1322.

[6] S.M. Kay, *Fundamentals of Statistical Signal Processing: Estimation Theory,* Prentice Hall, 1993.

[7] T. Moser, *Geophysics,* 1991, vol. 56(1), pp. 59-67.

[8] C. Holmes, B. Drinkwater and P. Wilcox, *NDE E Int.,* 2005, vol. 38, 701-711.

In: Ultrasonics: Theory, Techniques and Practical Applications ISBN: 978-1-62257-685-2
Editors: Hanako Ayabito and Mitsuko Katsukawa

Chapter 8

Application of Ultrasonics as Assistance in Machining Operations

A. Celaya**, L. N. López de Lacalle, F. J. Campa and A. Lamikiz
Department of Mechanical Engineering – University
of the Basque Country Faculty of Engineering of Bilbao, Bilbao, Spain

Abstract

The objective of the vibration assisted machining techniques is to improve the traditional cutting processes by the application of an external vibration to the cutting edge to achieve improvement in the chip removal mechanism and in the manufactured part quality. Currently, two techniques are most common for vibration-assisted machining due to their feasibility and industrial expectations: the Modulation-Assisted Machining and the Ultrasonic Assisted Machining.

Modulation-Assisted Machining superimposes a controlled low-frequency oscillation to the cutting process, typically up to 1000Hz with amplitudes up to 500 micrometers. The mechanics of the chip formation is changed in that way, and cutting is divided into discrete events by means of control of the modulation parameters. Thus, it improves efficiency and process capability in precision machining processes.

On the other hand, Ultrasonic Assisted Machining consists in adding high frequency excitation, which can reach 40 KHz, and low vibration amplitude, ranging between 3 and 30 micrometers, to the tool or workpiece to improve the cutting process. Ultrasonic assisted machining technology demonstrates a range of benefits in both i) high precision machining applications and ii) machining of difficult-to-cut alloys and brittle materials. Ultrasonic assistance is applied to several cutting processes, but most of the research on this method is focused on turning and drilling processes.

In this chapter, we report the basis of these techniques and the advantages and drawbacks of ultrasonic assisted turning. Thus, the effect of tool vibration on cutting process and workpiece surface quality was studied in turning of mild steel. Several parameters of ultrasonic assisted turning (UAT) were monitored, including surface roughness, tool wear, chip formation and tool temperature. The results obtained show that ultrasonic vibration can improve the surface quality decreasing surface roughness and

* Corresponding author, Email: ainhoa.celaya@bi.ehu.es.

cutting tool temperature and also a tool life increase is achieved compared to conventional turning (CT).

1. INTRODUCTION

Vibration-assisted machining consists on introducing a tool vibration to improve a manufacturing process. It has been applied to a number of processes such as turning, drilling, grinding and milling. Two different techniques can be applied depending on the type of vibration applied to the cutting tool, the Modulation-Assisted Machining and Ultrasonic Assisted Machining.

The Modulation-Assisted Machining applies a controlled low-frequency modulation to the metal cutting process [M4S 2009] [MAN 2007]. The frequency ranges up to 1,000Hz with amplitudes up to 500 micrometers. The controlled modulation may be applied on the cutting tool or on the workpiece in the direction of tool-feed, cutting velocity, or any combination offering major beneficial effects on the mechanics of machining. Chip breakage in machining processes (e. g., turning, boring, drilling) was found to be a consequence of the chip thickness becoming zero during each cycle of modulation under certain modulation conditions. Good results are reported in drilling [CHH 2002] of metal, brittle materials [BRE 2008], and composites using this method.

The Ultrasonic Assisted Machining consists on applying a high frequency (16-40KHz) and low amplitude (3-30μm) vibration to the tool or workpiece in order to improve the manufacturing process. Ultrasonic assisted machining was first used in the 1960s [SKE 1969], but recently there has been a growing interest and tendency to employ this technique for both high precision machining applications and for difficult-to-cut material machining. Although this technology has demonstrated a range of benefits in machining difficult-to-cut alloys and brittle materials it is still not widely used in industry.

Ultrasonic assistance is applied to several manufacturing methods, but most of the research on this technique is focused on turning and drilling processes. Hence, this work will be focused on Ultrasonic Assisted Turning.

In a turning process there are three possible directions in which the ultrasonic vibration can be applied to the cutting tool, which are the cutting speed direction, feed direction and radial direction. Vibration of the cutting tool in a plane including the cutting direction and the chip flow direction, known as elliptical vibration, has also been studied. Elliptical vibration assisted turning was first presented in the 1990s [SHA 1994] and it is mainly used for precision machining applications with depths of cut below 50 μm [MOR 1995] [LI 2006].

Ultrasonic assisted turning has been widely studied and there are quite a lot of works that have proved the advantages of this technique in terms of cutting forces, surface finish and tool wear in several materials. Conventional machining of nickel- and titanium-based alloys presents high tool temperatures and fast cutting edge wear. High-frequency vibration of the cutting tool has reportedly allowed a significant reduction of cutting forces and tool wear, a surface finish improvement up to 25-40%, as well as roundness improvements up to 40-50% [BAB 2003]. When cutting low alloy steels (DF2), the ultrasonic vibration means a reduction in cutting forces up to 50% approximately, and produces comparatively smaller chips and improved surface finish compared to conventional turning [NAT 2007]. Precision turning of stainless steel has also been studied by applying ultrasonic vibration to the diamond tool

[MOR 1991] in order to reduce tool wear. In other studies [BAB 2003, CEL 2009] improvements in roughness are found in the case of vibration assisted turning of mild steels.

Machining of brittle materials is also another field of application for this technique. To produce high quality surfaces on brittle materials, it is important to achieve a ductile deformation mode. Ultrasonic assisted machining can increase the critical depth of cut at which ductile-cutting can be obtained. The tool wear was also reduced by applying ultrasonic vibration, comparing to conventional turning [ZHO 2006].

In short, there are several advantages of introducing an ultrasonic vibration in the tool: reduction of cutting forces, shorter chips, surface roughness improvement, tool wear reduction, roundness improvement, and, as a result, an increase of the machinability of difficult-to-cut materials.

Finally it shall be mentioned that ultrasonic assisted machining is different for techniques using only ultrasonics and abrasive powders, as the traditional ultrasonic machining (USM), in the market by Extrudehone, or rotary ultrasonic machining (RUM). In these cases there are not cutting tools with defined edge involved in the process and ultrasonic excitation produce the rapid movement of abrasive particles that remove material in each of the small and quick hits against materials. Abrasive powders are in suspension in an oil fluid and must be all time injected in the process area.

2. Ultrasonic-Assisted Turning

In ultrasonic-assisted turning (UAT), the vibration is applied to the cutting edge of the tool, in one of three possible directions: the cutting speed direction, the feed direction, and a radial direction. Nevertheless, for the ultrasonic vibration effect to become noticeable, the vibration velocity must be an order of magnitude higher than the velocity of cutting in the direction of the vibration [KIM 1997]. So, when the vibration is applied in the cutting direction, the vibration velocity $vt=2\pi \cdot a \cdot fr$ has to be higher than the cutting speed $v=\pi \cdot n \cdot d$. And, when the vibration is applied in the feed direction, the vibration velocity $vt=2\pi \cdot a \cdot fr$ has to be higher than the feed speed $f \cdot n$, where n is the workpiece rotation speed, d the workpiece diameter, f is the feed, a, the vibration amplitude and fr the vibration frequency. In this way, during each vibration cycle there is a separation between the workpiece and the tool as seen in Figure 1, which promotes a pumping action that decreases the friction [ZHO 2002]. Hence, the reduction in the cutting heat and the work material adherence to the cutting edge of the tool increases the accuracy of the obtained workpieces [MUR 1998, AST 2006].

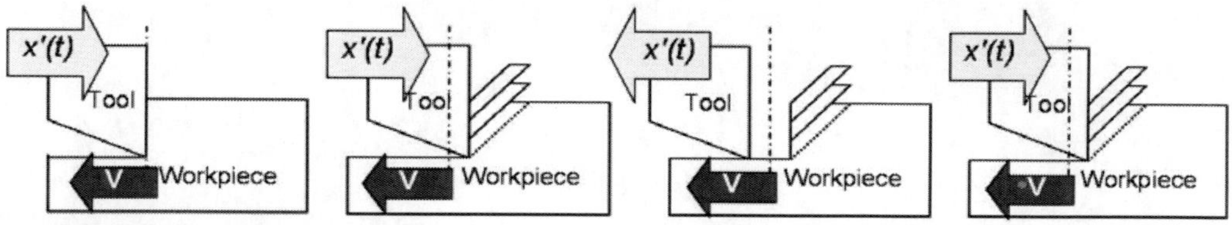

Figure 1. Principle of ultrasonic assisted machining.

In several works, when the vibration is applied in the cutting speed direction, the tested cutting speeds are lower that those applied in the usual industrial applications. The reason is that, for an ultrasonic vibration with amplitude of 10μm and a frequency of 20 kHz, the cutting speed has to be very low for conventional materials, about 75m/min in order to notice the effect of the ultrasonic vibration. However, when the vibration is applied in the feed direction the parameters used in manufacturing industry can be reached as, usually, feed speeds are below *vt*. Consequently, tool vibration in the feed speed direction seems to be more suitable for industrial ultrasonic turning applications requiring high productivity.

In this Chapter, first, an ultrasonic assisted vibration turning device is presented, a design implemented in a conventional lathe. Then, the device has been tested in order to investigate the effect of ultrasonic tool vibration on surface quality. Several turning tests have been done on AISI 1045 steel while changing the cutting speed, the feed and the depth of cut.

3. EXPERIMENTAL PROCEDURE

Cutting tests of the above setup were performed on an universal lathe. A commercial piezoelectric transducer Mastersonic MSG-2000 (range of frequencies: 17.5-24 kHz) was used to generate the ultrasonic vibration. A booster or sonotrode attached to the transducer was used to amplify the vibration. The design of the booster has to allow the transmission and amplification of the vibration generated in the transducer while maintaining enough stiffness to be able to cut the material without a high deflection or dynamic problems. For this purpose, the booster has two clamping rings designed by finite element technique to increase the system stability as shown in Figure 2. The obtained vibration at the tool tip has been measured with a CCD Laser Displacement Sensor.

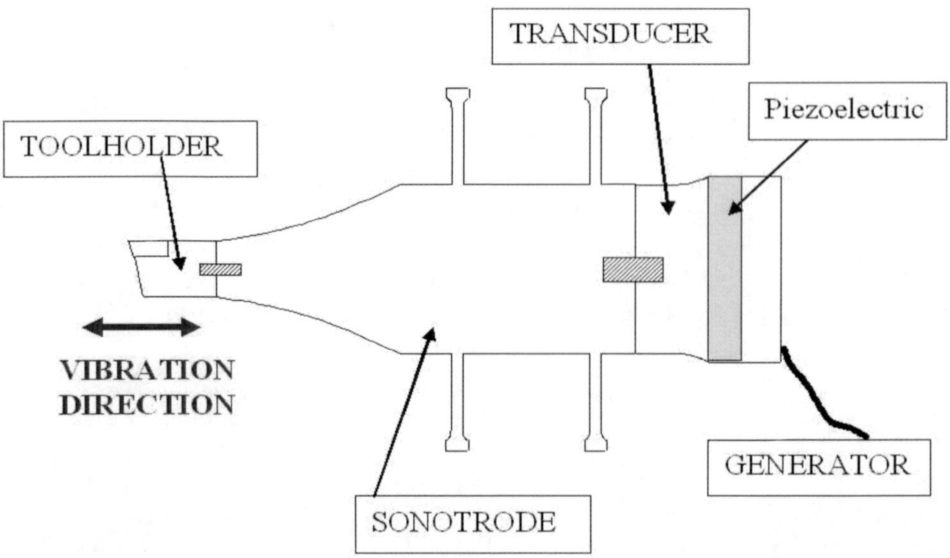

Figure 2. Designed booster.

Figure 3. Set up (a) orthogonal cutting and (b) straight turning.

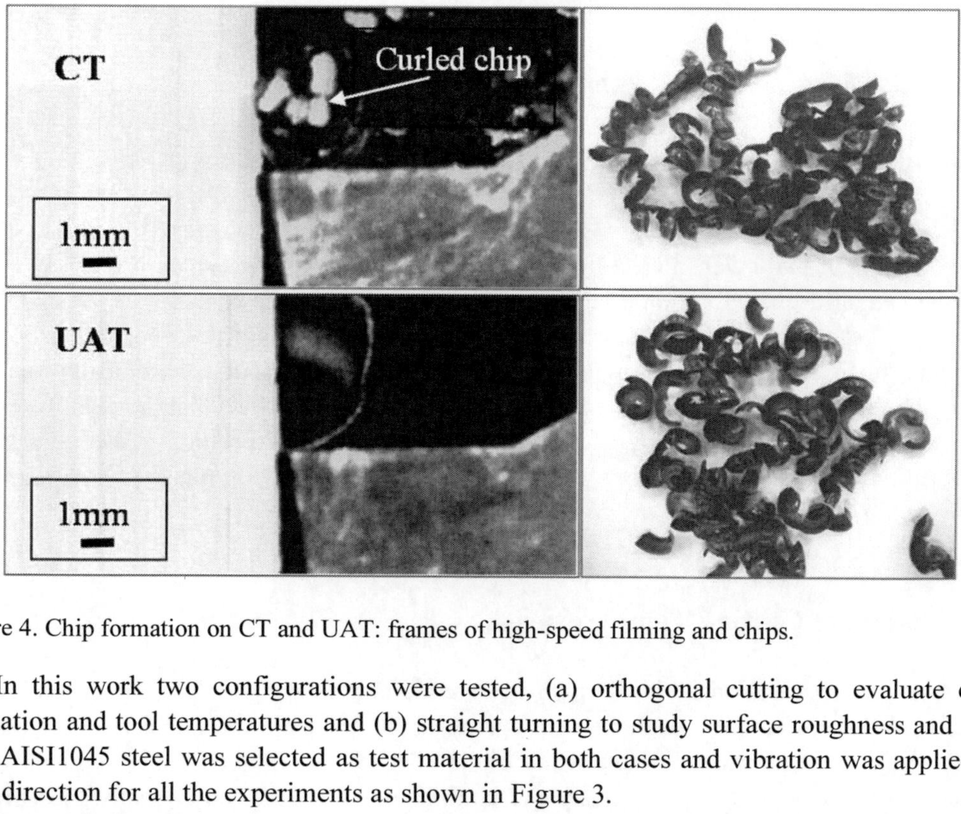

Figure 4. Chip formation on CT and UAT: frames of high-speed filming and chips.

In this work two configurations were tested, (a) orthogonal cutting to evaluate chip formation and tool temperatures and (b) straight turning to study surface roughness and tool life. AISI1045 steel was selected as test material in both cases and vibration was applied in feed direction for all the experiments as shown in Figure 3.

In orthogonal cutting, a grooving tungsten carbide tool (MITSUBISHI GY2M0600J040N-MS VP20RT) of 0° rake angle and 6° clearance angle has been used. In this configuration the resonant frequency was 19.5 KHz, with amplitude of 7 μm in the axial direction of the tool. To study chip formation, a high-speed camera (Photron Fast CAM-Ultima APX-RS 250K) was used for real-time observation. Filming speed was 50000 frames per second with an image area of 256x128. An infrared camera (Nikon Laird S270A) was also used for temperature distribution measurement in the cutting tool.

In straight turning, a tungsten carbide tool (Sandvik™ TNMG 110304-QM4225) was used with nose radius of 0.4mm, 6 rake angle and 6 clearance angle. The resonant frequency employed was 21.44 KHz, with amplitude of 6 μm in the axial direction of the tool. In the tests to study the effect of vibration assistance in the workpiece surface finish, all steel bars were machined prior to experiments to eliminate any eccentricity and to obtain the same initial conditions. The same cutting tests were carried out with and without vibration in order to compare the cutting performance between the two methods. The tests started with the application of ultrasonic vibration, and after the tool traversed 20mm, it continued without ultrasonic vibration under the same cutting conditions. Cutting conditions in the straight turning experiments were: depths of cut from 0.1 to 0.4mm; feeds from 0.05 to 0.2mm/rev; and cutting speeds from 75 to 125m/min. Surface quality was assessed by measuring the surface average roughness (Ra) and the mean roughness depth (Rz) along the axial direction of the workpieces. For the obtention of the roughness profiles, a Gaussian filter with a cut-off length of 0.8mm and a tracing length of 10mm was used. Both surfaces produced by ultrasonic assisted turning and conventional turning were evaluated using a Taylor Talysurf series 2 surface measuring instrument.

4. Results and Discussion

4.1. Analysis of Chip Formation

Chip breakage and control is a fundamental aspect when machining mild steels. Therefore, it is of particular interest the study of differences observed in the formation of the chip when an ultrasonic vibration is superposed to the turning process.

Figure 4 shows an image obtained with the high speed camera that shows the tool tip, workpiece and chip interaction in both conventional turning and ultrasonically assisted turning. High speed filming conducted in conventional turning reveals that the chip formed is longer and with a greater tendency to curl. It is noted that the superposition of the vibration to the tool promotes the chip breaking producing more regular and shorted chips, resulting in improved chip evacuation.

4.2. Analysis of Cutting Temperatures

Temperature evolution during turning is an important aspect to take into account when CT and UAT are compared. In machining operations the temperature increase in the tool is produced due to the plastic deformation and friction between workpiece and cutting tool. This temperature increment affects seriously the cutting tool life and consequently the surface roughness.

In order to measure the temperature distribution in the area of interaction between the workpiece and the tool tip the thermal images were taken every 1/10 seconds since the contact until after 1.7 seconds turning in both CT and UAT to compare the temperature rise produced in the tool.

Figure 5 shows the images and the temperature values obtained in both conventional and ultrasonically assisted turning. Experimental results show that the tool temperature increment is slightly lower when ultrasonic vibration is applied to the tool in the feed direction. This value varies according to cutting conditions but can be up to 15% lower when ultrasonic-assisted turning. However, to obtain this temperature reduction it is necessary to have proper vibration conditions as when the tool vibration is not correct, tool temperature can be increase due the own vibration energy.

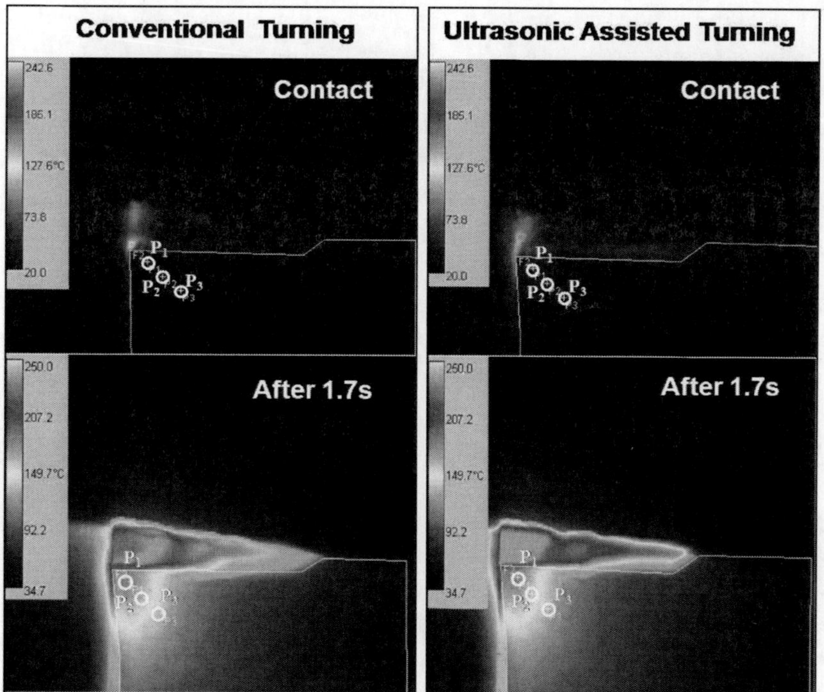

Conventional Turning

	TEMPERATURE INCREMENT (ºC)
P1	177,3
P2	160,8
P3	137,5

Ultrasonic Assisted Turning

	TEMPERATURE INCREMENT (ºC)
P1	161,1
P2	151,4
P3	128,7

Figure 5. Temperature distribution in the cutting tool.

4.3. Analysis of Surface Roughness

Figure 6 shows representative profiles of the machined surfaces of the steel parts measured in the axial direction. The profile above represents the conventional turning and the profile below belongs to the application of ultrasonic vibration. The roughness profiles have been represented on the same scale to facilitate comparison. The experimental results show that the fluctuations which occur in the surface roughness obtained in conventional turning are greater than those obtained in ultrasonic assisted turning. The reason for the reduction of fluctuations is that the machining process becomes more stable with the application of vibration at the tip of the tool.

Although to study the differences between conventional and assisted turning conventional amplitude parameters such as Ra and Rz can be used, for a more in-depth analysis of the surfaces, a spectral analysis of the profiles has been performed. These kind of techniques can be useful when analyzing repetitive or periodic processes such as turning or milling. For this purpose, the roughness profiles of Figure 6 have been used as sources for the Fast Fourier Transform (FFT) analysis shown in Figure 7.

The results shown in Figure 7 indicate differences between the two machined surface profiles. First, looking at the amplitude of the FFT, the peak corresponding to the rotation frequency has been reduced with UAT. That means that the crests of the profile left by the tool feed were decreased when the vibration was applied to the tool tip. The analysis also shows a reduction of the amplitude in the longer wavelength components when ultrasonic vibration is applied, which means that UAT results in a smoother profile. Finally, at wavelengths shorter than the feed per revolution, the amplitude of the spectrum is a bit lower in the case of UAT, although it is not easy to relate that effect directly with the ultrasonic vibration.

Figure 8 shows the influence of cutting speed on the surface roughness obtained in the machined parts both in conventional and ultrasonic assisted turning. In both cases it is observed that the surface finish slightly improves with increasing cutting speed. The surface finish improvement at higher cutting speeds is usual in the machining of metals, and this phenomenon is usually explained based on the built up edge. Formation of built up edge is favoured in a certain cutting speed range, thus increasing the cutting speed above this range, built up edge is removed and as a result the workpiece surface finish is improved. However, there are cases where even when being outside this cutting speed range and therefore not having built up edge, the roughness continues improving when cutting speed increases. It is therefore necessary to explain this phenomenon in more detail.

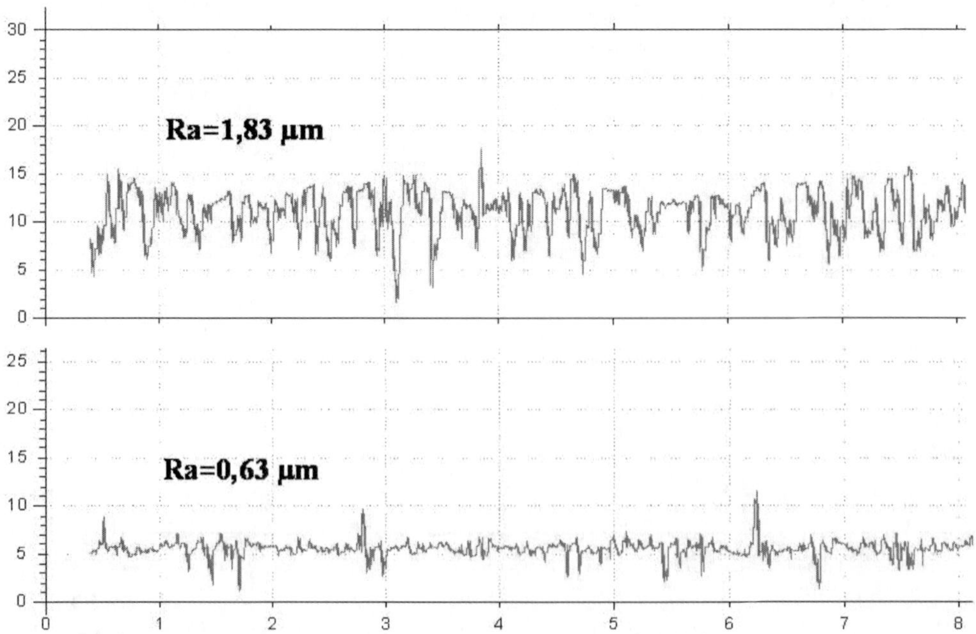

Figure 6. Roughness component of the machined surface: (a) without ultrasonic vibration; (b) with ultrasonic vibration (f=0.05mm/rev, ap=0.3mm, Vc=100m/min).

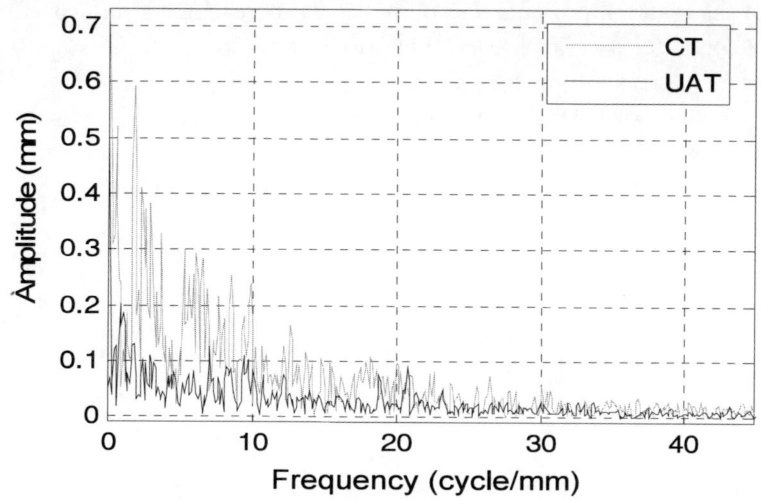

Figure 7. FFT diagram of the surface profiles of the conventionally machined (CT) and ultrasonically machined (UAT) workpieces(f=0.05mm/rev, ap=0.3mm, Vc=100m/min).

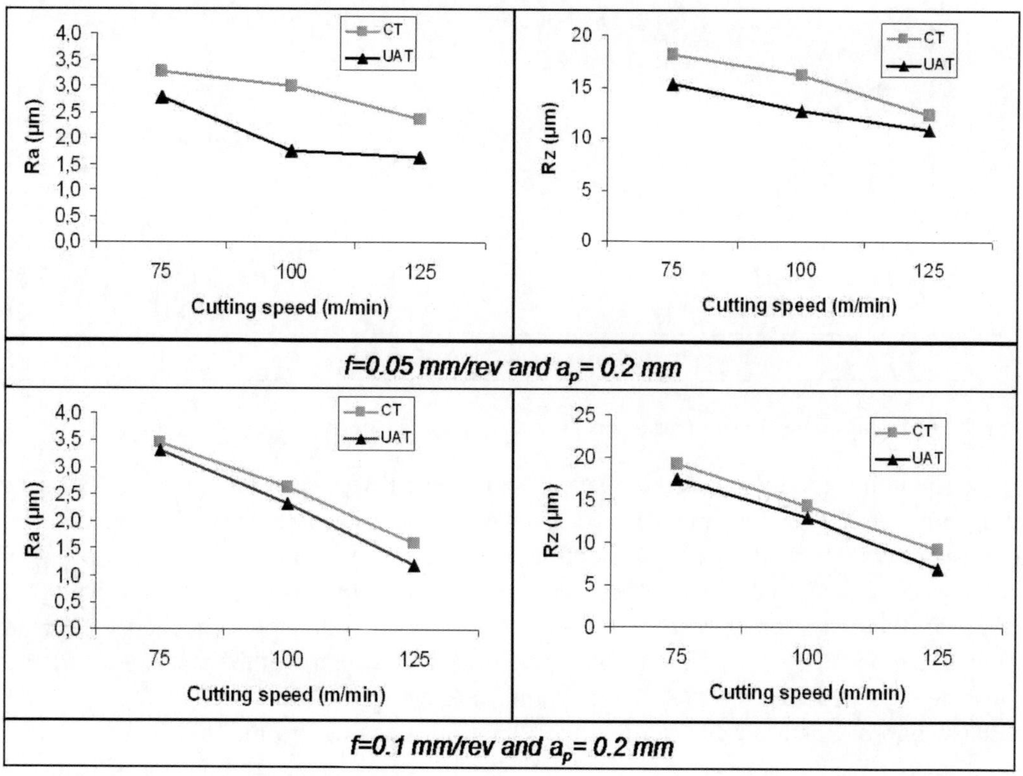

Figure 8. Cutting speed effect on surface roughness values.

Here, two possible reasons for this phenomenon [CHE 2000] are related. The properties of the metals are influenced by the deformation velocity, so the greater this velocity is, the less significant will be the plastic behaviour. Lateral plastic flow that occurs in the material of the workpiece along the direction of the edge may increase the peak-valley height of surface irregularities. If the material presents less plasticity by increasing cutting speed and therefore deformation velocity, surface finish can be improved as a result of a less lateral plastic flow and, consequently, less additional increase in the peak-height valley of the workpiece roughness. The second possible reason can be that when low cutting speeds are used, further grooves in the rake face appear. So, these defects of the tool are in part copied to the generated surface when the cutting edge comes into contact with the workpiece.

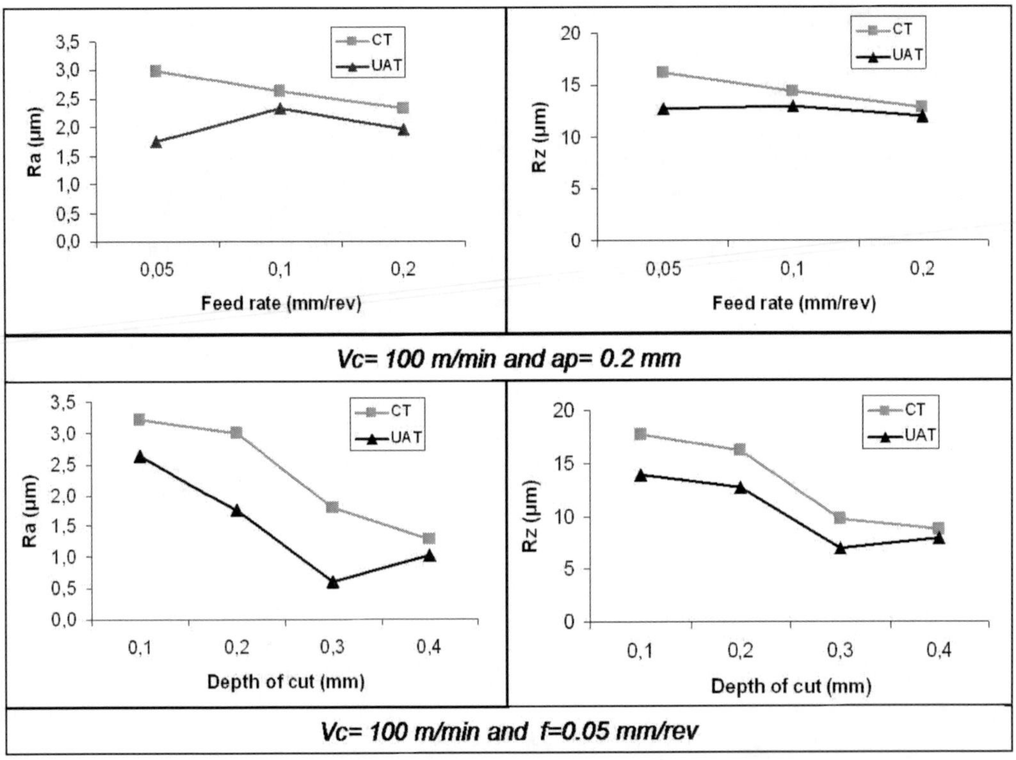

Figure 9. Feed rate and depth of cut effect on surface roughness values.

For all cutting conditions used in the tests, it is generally observed that ultrasonic assisted turning produces lower roughness values than the conventional turning. This phenomenon may be due to the workpiece-tool contact time reduction when vibration is applied to the tool in the feed direction, in this way defects of the rake face are less copied on the workpiece surface. Furthermore, the vibratory movement causes that the tool tip covers a greater workpiece surface at the same time reducing the peak-valley height of the machined surface.

In Figure 9 we can see that for the same cutting speed value and depth of cut, the roughness improvement produced when the vibration is applied to the tool is greater when using a smaller feed. But, in general, the roughness increases as the feed increases, being this trend less significant for large tip radius tools [CHE 2000]. So, the results obtained in Figure 9 initially contradict the classical roughness geometrical formula, but were caused because the applied cutting parameters were out of the optimal window application parameters given by the tool manufacturer. However, experiments with low values for both feed rate and depth of cut (i.e. a low chip section) were highly recommended because we try to improve the super-finishing operations. For example, when feed was 0.05 mm/rev, measured amplitude was 0,01mm, that is 1/5 ratio and therefore more significant effect of US was obtained as shown in Figure 9. For semi-finishing and roughing operation more powerful US transducer will be required. Anyway, in all range of cutting parameters the surface roughness using UAT decreases; in the best of the cases, it was improved over 60%.

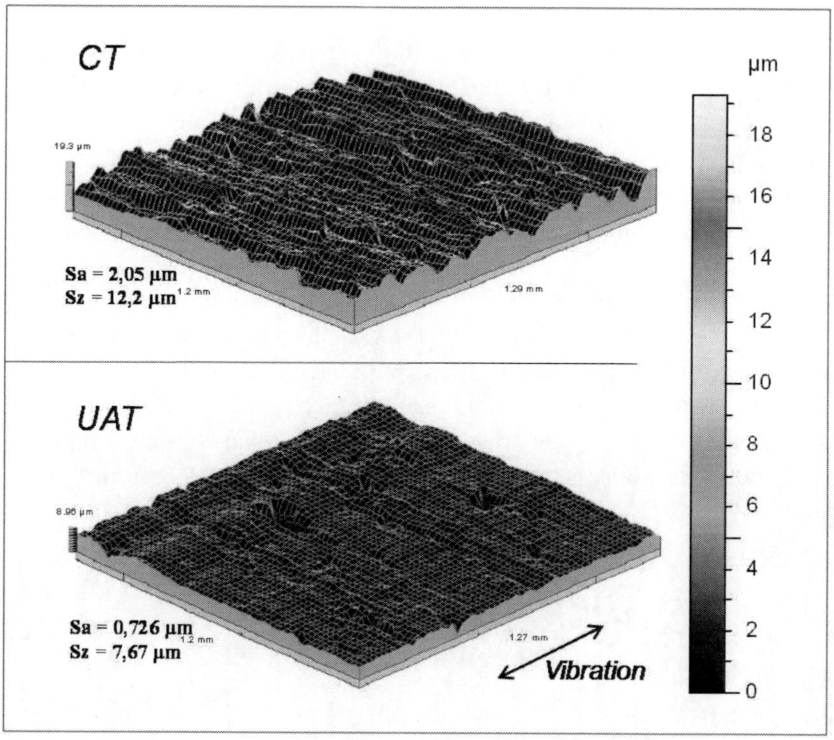

Figure 10. Surface topography for CT and UAT (f=0.05mm/rev, ap=0.3mm and Vc=100m/min).

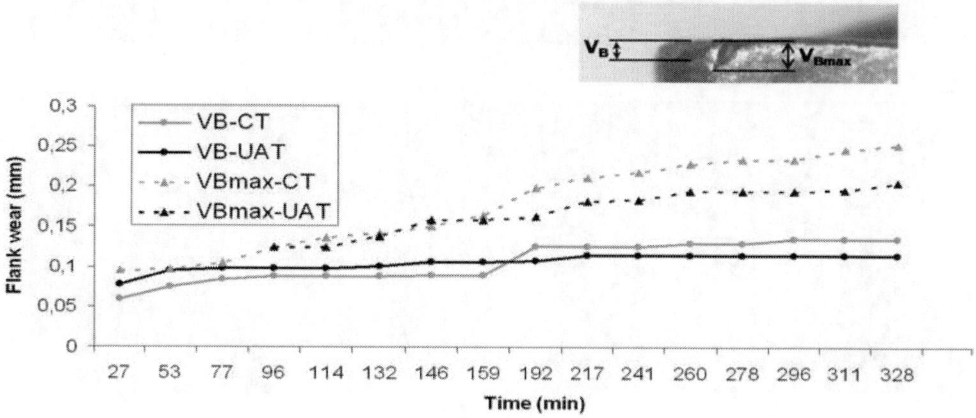

Figure 11. Flank wear VB and VBmax vs cutting time for CT and UAT. Cutting parameters: f=0.05mm/rev; Vc=100m/min; ap=0.3mm.

The surface topography in both conventional turning and vibration assisted turning is presented in Figure 10. UAT results show that the machined peaks are reduced. But, there are surface areas with higher peak amplitude, which can be produced by lack of uniformity in tool vibration. So, if a uniform tool vibration and good amplitude is used in UAT, the surface roughness can be improved.

4.4. Analysis of Tool Flank Wear

Figure 11 illustrates the VB and VBmax flank tool wear obtained during life tests for both CT and UAT. In both cases, a higher tool wear was produced in conventional turning compared with ultrasonic assisted turning. This effect can be explained because in conventional turning there is continuous friction at tool-workpiece interface and in assisted turning there is a separation between the workpiece and the tool. Since flank wear is mainly caused by abrasion, this implies that in UAT there is less abrasion and as a result reduced tool wear.

In Figure 11 can also be observed that tool wear increases more slowly in UAT than in CT after a certain time. This lower increase in UAT wear is due to the discontinuous cutting process. By contrast, in CT this wear increase is greater due to continuous cutting process, which causes a temperature increase and as result higher tool wear. Therefore, the ultrasonic assisted process show lower wear rates than conventional turning. Figure 12 shows photographs of tool wear after the end of the machining operation both in conventional and assisted turning.

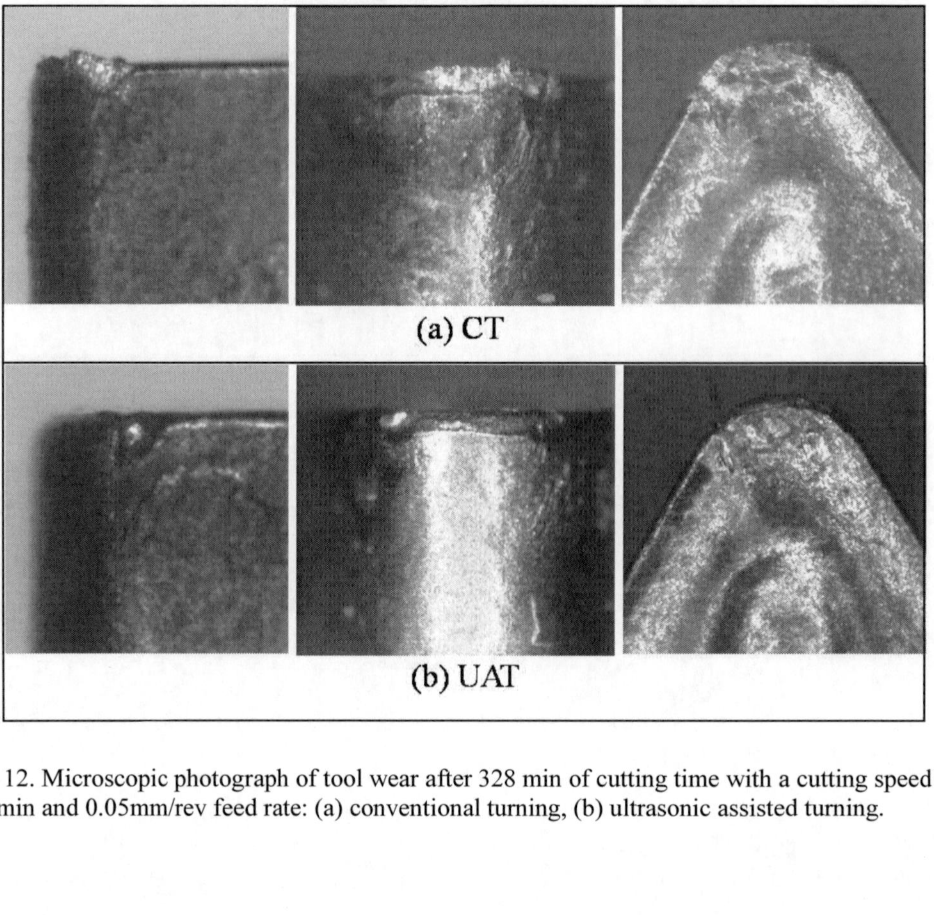

Figure 12. Microscopic photograph of tool wear after 328 min of cutting time with a cutting speed of 100m/min and 0.05mm/rev feed rate: (a) conventional turning, (b) ultrasonic assisted turning.

It should be noted that in both cases the tool life is extremely large, because of the low cutting parameters used in the experimental test for mild steel machining, usual for large batch production.

CONCLUSION

In this work, the effect of ultrasonic vibration in feed speed direction has been studied. For that purpose, an ultrasonic assisted vibration turning device has been implemented and tested on mild steel to investigate the effect of ultrasonic tool vibration on chip formation, surface quality, tool wear, and tool temperature.

The roughness of the turned steel surfaces with UAT has improved about 60% depending in cutting and vibration conditions. These roughness reductions depend on uniformity and amplitude value of vibration so, as chip section increases, a higher ultrasonic power is necessary. Also, it has been demonstrated through a spectral analysis that the vibration in the feed speed direction reduces the height of the crests produced by the tool and reduces the waviness of the generated geometry.

Chip formation has been observed by high-speed camera and it has been stated that better chip breakage is produced with vibration assistance.

Temperature measurements of the cutting tool have also shown that lower temperature increments are obtained when vibration is used. Also a tool flank wear reduction is observed in tools with UAT.

To sum up, experimental studies of conventional turning and ultrasonically assisted turning have demonstrated the advantages of cutting tool vibration assistance on mild steels machining.

ACKNOWLEDGMENTS

The authors acknowledge to CIC marGUNE, and to the department of industry of the Basque Government for its financial support. Special thanks to J.Pujana for the wise advice in the past and to researchers of CIC marGUNE for collaborations in the recent past in this field.

REFERENCES

Astakhov, V. (2006). Tribology of metal cutting, *Elsevier*.

Babitsky, V. I., Kalashnikov, A. N., Meadows, A., Wijesundara, A. A. H. P. (2003). Ultrasonically assisted turning of aviation materials, *Journal of Materials Processing Technology*, 132, 157-167.

Brehl, D. E., Dow, T. A., (2008), Review of vibration-assisted machining, *Precision Engineering*, 32, 153–172.

Celaya, A. , López de Lacalle, L. N., Campa F. J., Lamikiz A., Ultrasonic assisted turning of mild steels, *International Journal of Material and Product Technology*, Vol. 37, No.1/2, 60-70.

Chen W. (2000) Cutting forces and surface finish when machining medium hardness steel using CBN tools, *International Journal of Machine Tools & Manufacture*, Vol. 40, 455-466.

Chhabra P. N., Ackroyd B., Compton W. D., Chandrasekar S, (2002). Low-frequency modulation-assisted drilling using linear drives, *Proceedings of the Institution of Mechanical Engineers. Part B. Journal of engineering manufacture*, 216/ 3, 321-330

Kim, J-D., Choi, I-H. (1997). Micro surface phenomenon of ductile cutting in the ultrasonic vibration cutting of optical plastics, *Journal of Materials Processing Technology*, 68, 89-98.

Li, X., Zhang, D. (2006). Ultrasonic elliptical vibration transducer driven by single actuator and its application in precision cutting, *Journal of Materials Processing Technology*, 180, 91-95 M4 Sciences Corporation, available at http://www.m4sciences. com/news/.

Mann J. B., Chandrasekar S., Compton W. D., (2007) Tool holder assembly and method for modulation-assisted machining, PCT/US2007/000730.

Moriwaki, T., Shamoto, E. (1991) Ultraprecision diamong turning of stainless steel by applying ultrasonic vibration, *Annals of the CIRP, 40/1*, 559-562.

Moriwaki, T., Shamoto, E., (1995) Ultrasonic Elliptical Vibration Cutting, *Annals of the CIRP, 44/1*, 31-34.

Murakawa, M., Jin, M. (1998) Turning of beta-titanium alloys by means of ultrasonic vibration, *Trans. NAMRI/SME, 26*.

Nath, C., Rahman, M., Andrew, S.S.K. (2007) A study on ultrasonic vibration cutting of low alloy steel, *Journal of Materials Processing Technology*, 192–193, 159–165.

Shamoto, E., Moriwaki, T., (1999) Ultraprecision Diamond Cutting of Hardened Steel by Applying Elliptical Vibration Cutting, *Annals of the CIRP*, *48/1*, 441-444.

Skelton, R.C., (1969) Effect of Ultrasonic Vibration on the Turning Process, *Int.J.Mach. Tool Des. Res.*, 9, 363-374.

Zhou, M., Wang, X. J., Ngoi, B. K. A., Gan, J. G. K. (2002) Brittle-ductile transition in the diamond cutting of glasses with the aid of ultrasonic vibration, *Journal of Materials Processing Technology*,121, 243-251.

Zhou, M., Ngoi, B. K. A., Yusoff, M. N., Wang, X. J. (2006) Tool wear and surface finish in diamond cutting of optical glass, *Journal of Materials Processing Technology*, *Vol.* 174, 29-33.

INDEX

A

B

C

D

E

F

G

H

I

J

K

L

M

N

O

P

W

X

Y